Walery Jaworski, Albert L. A. Toboldt

The Action, Therapeutic Value and Use of the Carlsbad Sprudel Salt

powder form

Walery Jaworski, Albert L. A. Toboldt

The Action, Therapeutic Value and Use of the Carlsbad Sprudel Salt
powder form

ISBN/EAN: 9783337410438

Printed in Europe, USA, Canada, Australia, Japan

Cover: Foto ©berggeist007 / pixelio.de

More available books at **www.hansebooks.com**

THE

ACTION, THERAPEUTIC VALUE AND USE

OF THE

CARLSBAD SPRUDEL SALT

(POWDER FORM)

AND ITS RELATION TO THE

CARLSBAD THERMAL WATER

BY

Dr. W. JAWORSKI,

DEMONSTRATOR AT THE UNIVERSITY OF KRAKOW.

CLINICAL EXPERIMENTAL RESEARCHES MADE AT THE UNIVERSITY CLINIC OF
PROF. KORCZYNSKI IN KRAKOW.

WITH A

DIETARY

BY THE TRANSLATOR,

A. L. A. TOBOLDT, M.D.,

ASSISTANT DEMONSTRATOR OF PHARMACY, UNIVERSITY OF PENNSYLVANIA;
EDITOR JOURNAL OF BALNEOLOGY AND MEDICAL CLIPPINGS, ETC.

PHILADELPHIA:
P. BLAKISTON, SON & CO.,
1012 WALNUT STREET.
1891.

PREFACE.

The increasing interest manifested in this country in natural mineral waters and in products derived therefrom, coupled with the almost total lack of really scientific research in this branch of medicine, have been the incentives that have led to the translation of this work from the German of the renowned balneologist, Dr. W. Jaworski.

As the Carlsbad Thermal Waters have for over 200 years held such a prominent place in the treatment of disease in Europe, and as American physicians are becoming more and more impressed with the importance of making themselves familiar with the therapeutic value of these as well as other mineral springs, a systematic and scientific treatise, together with carefully tabulated experiments, must of necessity prove acceptable.

Works without number have preceded the present one, and very many of them have been carefully examined by the translator. Dr. W. Jaworski has improved on his predecessors in not contenting himself with experiments on animals, but has himself made a long series of most careful clinical experiments on human subjects. His researches are therefore entitled to much greater weight and consideration than those previously recorded.

THE TRANSLATOR.

CONTENTS.

	PAGE
SECTION 1. PREFACE,	9

I—EXPERIMENTS TO DETERMINE THE INFLUENCE OF CARLSBAD SPRUDEL SALT ON THE APPEARANCE OF THE STOOLS.

SECTION 2. ANNOTATION OF EXPERIMENTS, ... 12
    Experimental Series I, ... 13
SECTION 3. GROUPING OF THE RESULTS FROM EXPERIMENTAL SERIES I, ... 22
SECTION 4. SPECIAL EXPERIMENTS MADE TO DETERMINE THE NATURE OF THE EVACUATIONS UNDER THE INFLUENCE OF THE SPRUDEL SALT, . . 26
SECTION 5. RÉSUMÉ OF THE EFFECT ON THE EVACUATION OF THE BOWELS AFTER THE ADMINISTRATION OF SPRUDEL SALT, ... 28
SECTION 6. THE ACTION OF THE SOLUTION OF SPRUDEL SALT ADMINISTERED BY THE RECTUM, ... 29
SECTION 7. RÉSUMÉ OF THE ACTION OF THE SPRUDEL SALT USED BY INJECTION PER ANUM, ... 31

II—EXPERIMENTS MADE WITH CARLSBAD SPRUDEL SALT (POWDER FORM) ON THE FUNCTIONS OF THE STOMACH.

SECTION 8. DESCRIPTION OF THE METHOD OF EXPERIMENTING, ... 32
    Table of Experiments.
SECTION 9. DEDUCTIONS FROM THE TABLE OF EXPERIMENTS, ... 50
SECTION 10. THE EFFECT OF ADDITIONAL DOSES OF SPRUDEL SALT UPON THE FUNCTIONS OF THE STOMACH, ... 57
SECTION 11. CHANGES IN THE ENTIRE GASTRO-INTESTINAL FUNCTIONS AFTER CONTINUED USE OF THE SPRUDEL SALT, ... 58
SECTION 12. THE TOTAL RESULT OF THE INFLUENCE OF SPRUDEL SALT ON THE ENTIRE GASTRO-INTESTINAL FUNCTION, ... 64

III—CLINICAL DEDUCTIONS BASED UPON THE EXPERIMENTS MADE.

SECTION 13. THE RESTRICTIONS AND CONTRA-INDICATIONS IN THE USE OF THE CARLSBAD SPRUDEL SALT, ... 67
SECTION 14. INDICATIONS FOR THE USE OF SPRUDEL SALT, ... 69

MODE OF USE OF THE SPRUDEL SALT (POWDER FORM).

SECTION 15. APPLICATION OF THE SPRUDEL SALT BY THE MOUTH, ... 73

SECTION 16. THE DIETETIC RÉGIME DURING A TREATMENT WITH SPRUDEL SALT, 75

SECTION 17. USE OF THE SPRUDEL SALT IN WASHING OUT THE STOMACH—APPARATUS FOR GASTRIC IRRIGATION AND ASPIRATION, ... 77

SECTION 18. USE OF THE SPRUDEL SALT PER RECTUM. 83

SECTION 19. LENGTH OF TIME AND DIRECTING OF THE SPRUDEL SALT TREATMENT, 84
Concluding Remarks, ... 86

APPENDIX.

SECTION 20. THE RELATION OF THE SPRUDEL SALT TO THE CARLSBAD THERMAL WATERS, ... 87

SECTION 21. THE ACTION OF THE THERMAL WATER AND SPRUDEL SALT WHEN USED JOINTLY, ... 90

SECTION 22. EXPERIMENTS TO DETERMINE THE INFLUENCE OF EXERCISE UPON THE BEHAVIOR OF THE THERMAL WATER AND SPRUDEL SALT SOLUTION IN THE STOMACH, ... 92

DIETARY, ... 95

THE ACTION, THERAPEUTIC VALUE AND USE

OF THE

CARLSBAD SPRUDEL SALT

(POWDER FORM),

AND

ITS RELATION TO THE CARLSBAD THERMAL WATER.

SECTION I.

PREFACE.

Already in a work entitled "Keyserlichen Baad-Medici," by Strobelberger, published in 1630, a remedy is mentioned by the name "unser warmes Baadsalz," which was added to the Carlsbad Thermal Waters to promote the action of the bowels. The chemical composition of this salt cannot be positively determined, as the manner in which it was prepared is not given. It is only in the inaugural dissertation of Friedrich Hoffman (1734), that mention is made of the fact that in the preparation of the salt the Sprudelwasser was evaporated, filtered, cooled and allowed to crystallize. The remaining mother-liquor was repeatedly evaporated and cooled until no more crystals were obtained. At that time the salt was obtained only in small quantities. The "Brunnensalz" was obtained in larger quantities in 1764, and at the suggestion of the experienced and very astute Carlsbad physician, Dr. Becher, prepared in an essentially modified manner.

The natural heat of the spring was utilized and the water of the Sprudel evaporated on the spot, and thereby several crystallizations obtained. The crystals thus obtained were repeatedly redissolved in ordinary water, and again crystallized until they ceased to react alkaline. This "Brunnensalz" of Becher was therefore pure Glauber's salt. This mode of preparation was not adhered to so strictly later on, and the purification was not carried so far as Becher had carried it, or the same was omitted entirely; thus it came that until quite recently a product was brought into the market that was either pure Glauber's salt or else contained small quantities of sodium carbonate and sodium chloride amounting in the most favorable cases to only a few per cent. And it remains a puzzle why this coarsely crystallized salt, which appeared, until recently, in four-cornered bottles under the name of "Sprudelsalz," should for more than one hundred years be preferred to the cheaper, ordinary Glauber's salt.

After the chemist Göttl, in 1870, succeeded in conducting the Sprudel water, without changing its composition and without clogging the pipes, to any desirable distance, a very large and commodious municipal building was erected for the preparation of the Carlsbad salt in 1878–1879. It was as late as 1880, however, that the city council of Carlsbad decided, at a meeting specially convened for the purpose, to obtain a product which, in its composition, would be as near identical with that of the Sprudel water as possible. According to the findings and works of Prof. Ludwig (Ueber das Sprudelsalz, *Wr. Med. Blätter*, 1881, No. 1 bis 5) it was shown that such a product could be obtained by partly evaporating and filtering the Sprudel water; the filtrate to be evaporated almost to dryness and exposed to the action of the carbonic acid gas. According to these directions a new Carlsbad product has been produced, for the past four years, in the municipal establishment, under the exemplary direction of the city chemist, Dr. Sipöcz, whose well-known scientific and technical training and researches afford a guarantee for the accurate and scientific accomplishment. The mode of procedure is as follows: The Sprudel water is boiled, the sediment formed is removed by filtration, the filtrate is evaporated until a salt is obtained which still contains a certain per cent. of water, whereupon it is exposed to the action of the carbonic acid of the Sprudel spring itself, until it becomes saturated. The product thus obtained forms a white, finely crystalline powder, which is sent out into commerce in cylindrical bottles under the name of "Natürliches Karlsbader Sprudelsalz (pulver form)." The chemical composition of the same, according to repeated analyses of Dr. Sipöcz, is as follows:—

Sodium sulphate,	43.25 per cent.
Sodium hydrocarbonate,	36.29 "
Sodium chloride,	16.81 "
Potassium sulphate,	3.06 "
Lithium hydrocarbonate,	0.39 "
Sodium fluorate,	0.09 "
Sodium borate,	0.07 "
Silicic acid anhydride,	0.03 "
Iron oxide,	0.01 "

I have myself repeatedly analyzed the Sprudel Salt as found in commerce, and found the quantitative deviations of the main constituents to vary at the most but 2 per cent. I found, for instance, by trituration with normal HCl, the percentage of sodium bicarbonate to be 33.5 per cent. to 36.0 per cent.; with barium chloride the percentage of sodium sulphate 46.0 per cent. to 47.5 per cent.; with silver nitrate the percentage of sodium chloride 16.0 per cent. to 17.0 per cent. I have also always found the desired result, based on analysis by weight, that all carbonates exist as acid salts in the Sprudel Salt (powder form), which goes to prove that the same has been prepared conscientiously and carefully. The natural Carlsbad Sprudel Salt (powder form) therefore contains, with the exception of the silicic acid, aluminium oxide and the carbonates of calcium, magnesium, of iron and manganese, all of the soluble constituents of the Sprudel water, in the same combinations and quantitative relation in which they exist in the Sprudel water. I will treat later on in the chapter on the relation of the Sprudel Salt (powder form) to the Sprudel water. According to its chemical composition, the Sprudel Salt (powder form) is therefore a very desirable saline medicine, and it remains incomprehensible why, in the practice of some physicians, the crystalline should still be preferred to the Sprudel Salt (powder form), so that, at present, more must still be made of the former than of the latter. Inasmuch as I had found slight variations in the chemical composition of the Sprudel Salt, variations that were unavoidable, it became a matter of importance to me to obtain a homogeneous product of identical chemical composition for my clinical experiments. To obtain such a one I applied direct to Dr. Sipöcz, director of the municipal works, who, with the obliging consent of the Mayor, Herr E. Knoll, not only placed such a one at my disposal, but instituted a special quantitative analysis of the same. According to the analysis, the preparation used in my experiments was found to be of the following composition in relation to the main constituents: Sulphates, calculated as sodium sulphate, 46.56 per cent.; sodium chloride, 17.45 per cent.; sodium hydrocarbonate, 35.70 per cent.

According to the analysis of Prof. Ludwig and Prof. J. Mauthner, the dry residue for the Sprudel water was calculated as being 5.5168 grm. per litre. As during the evaporation the silicic acid, the carbonates of manganese, iron, calcium and magnesium are separated as an insoluble residue, 4.9527 grm. soluble constituents remain in Sprudel water which forms a corresponding quantity of Sprudel Salt (powder form). This quantity of Sprudel Salt dissolved in one litre of distilled water was to furnish one litre of artificial Sprudel water, without the above mentioned precipitated insoluble combinations. With the Sprudel Salt of the above composition I have instituted a series of experiments on the clinical patients of the medical clinic of Prof. Korczynski, in Cracow. The same afforded information on the following points:—

1. At what time after the ingestion of the Sprudel Salt (powder form),

in what number and with what accompanying phenomena do the stools appear?

2. What is the nature of the succeeding stools?

3. What change do the several constituents of the Sprudel Salt experience in the intestinal canal?

4. What doses, what concentration, and temperature of the Sprudel Salt solution are the most proper to act upon the intestine with as much energy as possible?

5. What is the action of the Sprudel Salt solution when introduced in the rectum?

6. In what manner is the acid gastric secretion influenced by the Sprudel Salt when taken once and when continued for a longer time?

7. What influence does it exert upon the digestive ferment and the digestive power of the gastric juice?

8. What is the condition of the mechanical capability of the stomach after continued use of the Sprudel Salt?

9. Is the action of the Sprudel Salt identical in healthy and diseased conditions?

10. May the action of the Sprudel Salt, after absorption, be observed on other organs than the stomach and intestinal canal?

11. In what relation does the Sprudel Salt stand to the Carlsbad water, *i. e.*, is their action upon the stomach and intestinal functions identical?

I.—EXPERIMENTS TO DETERMINE THE INFLUENCE OF THE CARLSBAD SPRUDEL SALT ON THE APPEARANCE OF THE STOOLS.

SECTION II.

ANNOTATION OF EXPERIMENTS.

These experiments were performed in the following way:—

The persons experimented upon took early, fasting, one or more normal doses of Carlsbad Sprudel Salt at intervals of $\frac{1}{4}$ to $\frac{1}{2}$ an hour. Thereupon the time when the first and the succeeding passages appeared was noted. A normal dose was considered to consist of 5 grm. Sprudel Salt dissolved in 250 c. c. distilled or ordinary water. The solution was drank either at the ordinary temperature of a living room, 18 to 20° C., or heated to 50° C.

On account of brevity and to afford an oversight, the observations obtained are arranged in Experimental Series:—

I. The appearance of the stools under the action of the Sprudel Salt in 34 individuals is there recorded. To these we add 3 other cases from Section 4, and 3 from Section 11, in whom the passages, although for another object, were subjected to a still more careful examination, so that the experiments upon the influence of Sprudel Salt on the evacuations is based on 40 cases. Of the 34 in Series 1 there are 7 with normal action of the bowels, 7 with catarrhal affection of the bowels, 15 with habitual constipation—cause not precisely determined, the majority probably depending upon atonic intestinal weakness—1 case with severe hæmorrhoidal knots, and 4 cases showing a partial impermeability of the intestinal canal. Six other cases are found further treated of in Sections 4 and 3.

EXPERIMENTAL SERIES I.

I. L. W., æt. 20, artisan; healthy, without any intestinal complaints. Stools regular and compact.

Experiment 1.—(*a*) The last passage early in the night before the experiment. Early, fasting, 5 grm. of Sprudel Salt in 250 c. c. distilled water at 18° C., were taken. No passage until early the next day, of the usual consistence and without special symptoms.

Experiment 1.—(*b*) 5 grm. Sprudel Salt dissolved in 250 c. c. artificial Seltzer water also remained without effect.

Experiment 2.—The last stool early in the morning before the experiment. 5 grm. Sprudel Salt in 250 c. c. of distilled water at 18° C., were taken early, fasting; after half an hour the same dose was repeated; not until evening (after 12 hours) did a passage, without special effort, follow.

II. P. L., æt. 22, farmer. Catarrhal gastric symptoms. Passages regular every day.

Experiment 3.—(*a*) 5 grm. Sprudel Salt were without effect on the passages.

Experiment 3.—(*b*) Last passage on the day before the trial. 5 grm. Sprudel Salt were taken dissolved in 250 c. c. of distilled water early, fasting; the same dose was repeated after half an hour. After 3½ hours compact and pappy masses in the stool without special effort. There was no second passage that day.

Experiment 4. Last passage on the day before the trial. The same directions as in 3 (*b*), with the exception that the Sprudel Salt solution was taken heated to 55° C. After 3½ hours a passage composed of compact and pappy masses, accompanied with a feeling of fullness and oppression in the abdomen.

III. R. J., æt. 30, tradesman. Nervous dyspeptic symptoms. Passages usually regular, with tendency to diarrhœa.

Experiment 5.—Last passage on the day before the trial. 5 grm. Sprudel Salt dissolved in 250 c. c. were taken fasting, at 20° C., followed by a walk. After 2 hours a pappy stool without special effort, after 3 hours a second similar passage.

Experiment 6.—On the day following the last trial, 5 grm. Sprudel Salt dissolved in 250 c. c. of distilled water at 20° C., were taken. After half an hour the same dose repeated, thereupon moderate exercise in the garden. After 1 hour a pappy stool, after 1½ hours a second, followed during the afternoon by two watery passages with sharp pains.

Experiment 7.—Last passage the day before. Early, fasting, 10 grm. Sprudel Salt in 250 c. c. distilled water at 18° C. were introduced by means of the stomach tube. After 2 hours a pappy stool; in the afternoon, two more watery passages without inconvenience.

Experiment 8.—Last passage early before the trial. Exactly the same as in 7, with the exception that the solution was administered heated to 55° C. After 2 hours a pappy stool; in the afternoon three watery passages accompanied with pain.

IV. M. M., æt. 24, farmer. Dilatation of the stomach, with symptoms of a partial impermeability of the large intestine. Passages regular every day.

Experiment 9.—Last passage on the day before the trial. Early, fasting, 5 grm. Sprudel Salt in 250 c. c. distilled water at 18° C., were taken. An ordinary passage during the afternoon.

Experiment 10.—Last passage the day before. Early, fasting, 5 grm. Sprudel Salt in 250 c. c. distilled water at 18° C., were taken. After half an hour the same dose was repeated. In this, as also in a number of other cases, colicky pains would follow the ingestion of the Sprudel Salt; the bowels, however, would not be moved until the afternoon, and be of ordinary consistence.

V. P. S., æt. 32, military officer. Hypochondriacal mood and constipation, in consequence of atonic intestinal weakness. Stool hard, not daily, irregular.

Experiment 11.—(*a*) 5 grm. Sprudel Salt in 250 c. c. distilled water invariably remained without effect.

Experiment 11.—(*b*) 10 grm. Sprudel Salt in a single as well as divided dose remained without effect. Only nausea was produced. For the purpose of evacuating the bowels Sprudel Salt solution was therefore given per rectum.

VI. K. L., æt. 25, country girl. Nervous dyspeptic symptoms with atonic intestinal weakness. Passages retained for several days.

Experiment 12.—5 grm. Sprudel Salt in 250 c. c. distilled water at 18° C., were without result in repeated trials.

Experiment 13.—Last passage three days ago. Early, fasting, 5 grm. Sprudel Salt in 250 c. c. of distilled water were taken, and repeated in half an hour. In many similar trials only nausea and "boring" pains in the abdomen were produced, but no passage. Sprudel Salt in solution was then given in the form of a clyster.

VII. S. T., æt. 22, shoemaker. Hypochondriacal mood, atonic intestinal weakness; passages retained for days.

Experiment 14.—Repeated use of 5 grm. Sprudel Salt in 250 c. c. of distilled or Seltzer water at 18 or 55° C., only caused weight and other unpleasant sensations in the abdomen, but no passage. Solution of Sprudel Salt was used per rectum.

VIII. G. W., æt. 45, laborer. Carcinoma and dilatation of the stomach. Atonic intestinal weakness. No passages for four to five days.

Experiment 15.—5 as also 10 grm. Sprudel Salt in 250 c. c. distilled water at 18° C., caused nausea, but no passage.

Experiment 16.—Last passage 5 days ago. Early, fasting, 15 grm. Sprudel Salt in 250 c. c. of distilled water at 18° C., were introduced through the stomach tube. After 7 hours a passage, composed of a hard and pappy mass.

Experiment 17.—Last passage 4 days ago. Early, fasting, 15 grm. of Sprudel Salt in 250 c. c. distilled water at 18° C., introduced through the stomach tube. After 6½ hours a hard, after 9 hours a second watery passage.

IX. H. Z., æt. 28, a seamstress. Atonic intestinal weakness. No passage for several days.

Experiment 18.—Last passage 3 days ago. Early, fasting, 15 grm. Sprudel Salt in spring water at 18° C., was administered in three doses at intervals of half an hour. Nausea, but no passage was produced. Two following similar trials produced the same negative result.

Experiment 19.—Last passage 3 days ago. Early, fasting, four 5 grm. doses Sprudel Salt, each dissolved in 250 c. c. distilled water at 18° C., were taken at intervals of half an hour. Slight nausea. After 4 hours the first, and after 4½ hours a second watery passage, without pain. In the afternoon two more watery evacuations, followed early the next day by a pappy stool.

X. Cz. K., æt. 26, accountant. Atonic intestinal weakness. No passages for 2 to 4 days.

Experiment 20.—(*a*) Twice repeated doses of 5 grm. Sprudel Salt remained without effect.

Experiment 20.—(*b*) Last passage 3 days ago. Early, fasting, 5 grm. Sprudel Salt in 250 c.c. spring water, at 16° C., were given and repeated in half an hour. After 20 minutes a rumbling and passing of wind; after 1¼ hours the first, and after 1¾ hours the second watery evacuation.

Experiment 21.—Last passage 4 days ago. The same direction, with the exception that the Sprudel Salt was dissolved in soda water. After 15 minutes rumbling and passing of wind, after three-fourths of an hour a hard, after 1½ hours a second pappy, and during the day 2 more watery evacuations.

XI. P. B., æt. 45, laundress. Insufficientia mitralis. Emphysema pulmonum. Obstipatio. No passage for several days.

Experiment 22.—Last passage 2 days ago. Early, fasting, 5 grm. Sprudel Salt in spring water at the ordinary temperature, were administered. After a few minutes nausea and retching set in. Difficulty of breathing and palpitation, but no passage. Two similar trials produced the same effect with reference to the symptoms. The Sprudel Salt, therefore, had to be discontinued.

XII. R. N., æt. 33, official. Nervous, sensitive individual. Passages irregular, with tendency to diarrhœa.

Experiment 23.—Last passage 3 days ago. Early, fasting, 5 grm. Sprudel Salt in 250 c.c. distilled water at the ordinary temperature, were taken. After 1½ hours a pappy; during the day 3 watery evacuations without pain. The headache from which the patient suffered disappeared after every administration of the Sprudel Salt.

Experiment 24.—Last passage the day before. The same dose, but dissolved in 250 c.c. soda water. After 1 hour and during the day 2 more evacuations.

In numerous other trials, 5 grm. Sprudel Salt were always sufficient to produce copious evacuations.

XIII. W. N., æt. 32, clerk. Atonic intestinal weakness. Bowels moved every second or third day.

Experiment 25.—Last passage 2 days ago. Dose of 5 grm. Sprudel Salt only caused a rumbling in the bowels, but no passage; the following was, therefore, at once tried on the next day.

Experiment 26.—Early, fasting, 5 grm. Sprudel Salt were taken, dissolved in 250 c.c. distilled water at 18° C., and repeated in half an hour. With rumbling and straining, a compact stool was produced after 1½ hours.

Experiment 27.—Last passage 2 days ago. The same as above, but dissolved in carbonic acid water. Fifteen minutes after the first dose rumbling and passing of wind; after an hour a compact, after 1½ hours and after 1¾ hours 2 watery, greenish and bad-smelling stools.

Experiment 28.—Last passage 2 days ago. The same proceeding, only that artificial Seltzer water was used as the solvent. After 1¼ hours rumbling and copious passage of flatus; after three-fourths of an hour a compact, and after 1, as also 1¼ hours, 2 more watery, ill-smelling, greenish passages.

Experiment 29.—Last passage 2 days ago. Early, fasting, 10 grm. Sprudel Salt dissolved in 500 c.c. distilled water at 18° C., were taken at one dose. The large quantity of liquid could only be swallowed with great effort; produced a bad taste, nausea and retching. Not until after 1½ hours did a compact, and a few minutes later a larger watery evacuation take place.

Experiment 30.—Last passage 2 days ago. Early, fasting, 5 grm. Sprudel Salt was taken in 250 c.c. distilled water at 18° C. This dose was repeated after half an hour, and in 1 hour thereafter a third dose was taken. After

1½ hours, accompanied with a great deal of rumbling, a pappy evacuation; after 1¾ and 2¼ hours watery stools with burning in the rectum. During the afternoon 2 more watery passages.

XIV. M. F., æt. 28, school-teacher. Atonic intestinal weakness. Passages retained 3 to 4 days.

Experiment 31.—Last passage 3 days ago. Early, fasting, 5 grm. Sprudel Salt were taken dissolved in 250 c.c. distilled water; after half an hour the same dose. A sensation of fullness in the abdomen was produced, with nausea and retching, but no passages.

Experiment 32.—Last passage 3 days ago. Early, fasting, 5 grm Sprudel Salt were taken in 250 c.c. distilled water at 18° C., and repeated in half an hour, to be followed by a third half an hour later. The first dose was speedily followed by nausea, the second by retching, and the third by vomiting. The stool, consisting of a compact pappy mass, appeared 6 hours later.

Experiment 33.—Last passage 3 days ago. Early, fasting, 5 grm. Sprudel Salt were taken in 250 c.c. soda water at the ordinary temperature; the same dose repeated after half an hour. Nausea and vomiting were produced, but no stool appeared on that day. The Sprudel Salt, therefore, had to be given by the rectum.

XV. P. O., æt. 54, clerk. Great atonic intestinal debility with hæmorrhoidal knots. The bowels can only be opened by means of injections.

Experiment 34.—Last passage by means of injections of water 2 days ago. Early, fasting, 5 grm. of Sprudel Salt were taken, dissolved in 250 c.c. rain-water at ordinary temperature. The same repeated after half an hour. Rumbling, a sense of fullness and straining was produced and prolapse of hæmorrhoidal knots, but no passage.

Experiment 35.—Last passage by means of injections of water 3 days ago. The same as above, with the difference that artificial Seltzer water was used as the solvent. After an hour, amidst severe straining, prolapse of a part of the rectum together with hæmorrhoidal knots, there appeared a stool consisting of a ball of feces tinged with mucus and blood. After 1½ and 1¾ hours this was followed by watery passages deeply colored with blood. In this case the Sprudel Salt was discontinued.

XVI. K. C., æt. 25, student. Catarrhal disease of stomach. Severe atonic intestinal weakness with hypochondriacal mood. Evacuation of the bowels is generally only accomplished by means of injections of water.

Experiment 36.—(*a*) Doses of 5 grm. Sprudel Salt, dissolved either in 250 c.c. of distilled or soda water, produced no stools. Only rumbling being produced.

Experiment 36.—(*b*) 10 grm. Sprudel Salt taken at one dose also produce no effect. Neither could the passages be expedited by exercise.

XVII. K. B., æt. 24, student. Catarrhal disease of stomach; atonic intestinal weakness. No passage for 3–4 days at a time.

Experiment 38.—Last passage 4 days ago. Early, fasting, 5 grm. Sprudel

Salt were taken in 250 c.c. distilled water at the ordinary temperature, the same dose repeated in half an hour. After 3 hours a passage consisting of balls of feces and a watery fluid; after 4, as also after 5 hours two more watery stools. The patient experienced a great sense of relief in the abdomen after the stools.

XVIII. W. K., æt. 26, clergyman. Catarrhal disease of stomach and bowels. Passages generally produced by means of injections of water.

Experiment 39.—Last stool 2 days ago, produced by means of injection of water. 2 hours after supper 5 grm. Sprudel Salt in 250 c.c. distilled water at 18° C. were taken; the same dose repeated in half an hour. During the night rumblings in the abdomen, sleep disturbed. Not until morning (after 12 hours) a mushy stool.

Experiment 40.—Last passage 3 days ago. 4 hours after dinner 5 grm. Sprudel Salt were taken dissolved in 250 c.c. of distilled water at 18° C.; the same dose repeated after half an hour. The patient complains of great distention of the stomach followed after three hours by a copious watery stool, accompanied by the passage of much flatus; during the night a second passage.

XIX. C. M., æt. 60, a functionary employed at the desk. Suffering from atonia intestinorum incipiens post enteritidem acutam. No passage for 2–3 days at a time.

Experiment 41.—For a long time 5 grm. Sprudel Salt was taken early, fasting, dissolved in 250 c.c. of soda water at the ordinary temperature. Regularly after 3–4 hours this was followed by a single passage without any sensation whatsoever. After 3 weeks' use of the Sprudel Salt the stools were regular.

XX. F. G., æt. 22, student. Catarrhal affection of stomach and bowels. Bowels not moved for 2–3 days.

Experiment 42.—In 5 trials, 5 grm. Sprudel Salt dissolved in 250 c.c. spring water and taken, would, after 1 to 1½ hours, produce 2–3 thin, watery passages without especial inconvenience.

XXI. Z. R., æt. 50, teacher. Catarrhus exacerbans intestini crassi. Evacuation of bowels only to be accomplished by the use of injections of water.

Experiment 43.—Last passage produced by means of injections of water 2 days ago. Early, fasting, 5 grm. Sprudel Salt in 250 c.c. of spring water at the ordinary temperature, and repeated in half an hour. After 1½ hours a compact stool, and after 1¾ hours a second watery passage mixed with blood, and accompanied with straining and a burning sensation in the rectum. The patient feels himself much exhausted during the remainder of the day.

Experiment 44.—Last passage 2 days ago. The same directions as above; as solvent, however, soda water was used. After 1 hour a pappy, mucous stool, consisting of hard balls of feces; after 1½ hours, at short intervals, numerous watery stools colored with blood, accompanied with a great deal of tenesmus and burning, so that the patient himself refused to continue the use of the Sprudel Salt.

XXII. M. A., æt. 28, drawing teacher. Catarrhal disease of the stomach. Symptoms of a partial obstruction of the intestine. Bowels irregular during a number of days.

Experiment 45.—Last passage 2 days before. 5 grm. of Carlsbad Sprudel Salt in 250 c.c. distilled water at 55° C., were taken early, fasting. After 6 hours a hard stool; early the following day a second, pappy passage.

Experiment 46.—Last passage 2 days before. 5 grm. Carlsbad Sprudel Salt in 250 c.c. of distilled water at 18° C., were taken fasting, early in the morning. In half an hour the same dose was repeated. After 2 hours nausea and vomiting set in, to be followed shortly by a large stool. During the night sharp pains were felt, confined to one spot; these were followed by 2 pappy passages. The following day the patient complained of general weakness. These symptoms supervened after every administration of 10 grm. of Carlsbad Sprudel Salt.

XXIII. K. P., æt. 22, laborer. Symptoms of dilatation of the stomach and stenosis of pylorus. Atonic intestinal debility. The bowels confined for 8–10 days at a time.

Experiment 47.—Bowels open 9 days before. 4 hours after dinner 15 grm. Carlsbad Sprudel Salt dissolved in 250 c.c. spring water at 17° C., were administered through the stomach tube. After 4 hours great flatulency, followed by nausea and pain in the abdomen. Not until 9 hours after, a copious passage, consisting of hard and pappy masses. On account of the distress in the abdomen the patient was restless and without sleep the whole night.

XXIV. S. T., æt. 39, cabinet maker. Excepting hæmicrania spastica, no evidence of any disease could be found. Passages daily, but with a tendency to constipation.

Experiment 48.—Last passage the day before. Early, fasting, 5 grm. of Sprudel Salt were taken, dissolved in spring water at the ordinary temperature. After 2 hours a pappy stool, without effort; no more passages that day. Cold and warm solutions in 5 trials produced identical results.

Experiment 49.—Last passage the day before. 5 grm. of Carlsbad Sprudel Salt were given early, fasting, dissolved in 250 c. c. of spring water at the ordinary temperature. After 2 hours a pappy and during the day another semi-liquid passage, without pain. The patient feels relief from his ailment during the day after this dose.

XXV. H. K., æt. 50, workman in factory. Well-nourished; atonia intestinalis, with dyspeptic troubles. Stools once every third day, hard and difficult to pass. Patient generally resorts to purgatives.

Experiment 50.—(*a*) Last passage the day before. Early, fasting, 5 grm. Carlsbad Sprudel Salt were taken in 250 c. c. spring water at 50° C., and repeated in half an hour. Immediately following the second dose, a firm passage, followed in half an hour by a pappy one, and another during the afternoon. The patient feels much better when taking the salt; no sense of weight in abdomen, better appetite and less stomach trouble.

Experiment 50.—(*b*) Last passage the day before. Early, fasting, 5 grm. of Carlsbad Sprudel Salt were taken, dissolved in 250 c. c. Mühlbrunnenwasser at 50° C.; this dose was repeated in half an hour. This was followed by a single passage without pain, 2 hours after the last dose; during other trials not until later in the day.

XXVI. P. L., æt. 40, single. Hysteria. Nervous individual. Every day or every other day a moderately hard passage.

Experiment 51.—Last passage 2 days before. 5 grm. Carlsbad Sprudel Salt were given early, fasting, dissolved in 250 c. c. distilled water. After 3 hours a pappy stool, and during the day 4 more liquid passages, with sharp pain in both hypochondriac regions. The patient feels "much relieved in the abdomen" on the days of trial.

Experiment 52.—Last passage the day before. 10 grm. Carlsbad Sprudel Salt in 250 c. c. distilled water at 50° C., were administered through the stomach tube. A pappy stool after fifteen minutes, and during the day a large number of watery passages, accompanied by continuous straining and pain in the hypochondriac region, followed by general exhaustion.

Even the dose of 5 grm. Carlsbad Sprudel Salt was in this individual too large.

XXVII. N. S., æt. 40, shoemaker. Catarrhus intestini cressi subs. obstipatione. Passages every 3 or 4 days, with difficulty.

Experiment 53.—Last passage 3 days ago. 10 grm. Carlsbad Sprudel Salt in 250 c.c. distilled water at 18° C., were taken early, fasting. After one hour a soft stool without pain, no more during the day.

XXVIII. S. A., æt. 35, peasant woman. Great dilatation of the stomach and constipation. In the stomach, whose capacity, as measured, was 6 litres, remnants of food could always be found early, fasting. According to her statement she would not have a passage for weeks at a time.

Experiment 54.—Last passage 10 days ago. Successive doses of Carlsbad Sprudel Salt to 20 grm. in 250 c. c. distilled water, administered through the stomach tube, never produce a passage, but severe pain in the stomach, distention in the region of the stomach and burning in the abdomen.

XXIX. S. C., æt. 45, farmer. Ectasia majoris gradus. Stenosis pylori post ulcus. Hæmatemesis. Would go weeks without having a passage. The capacity of the stomach, according to measurement, was 5 litres.

Experiment 55.—Last passage 7 days ago. Successive doses of Carlsbad Sprudel Salt to 20 grm. in 250 c.c. spring water, administered through the stomach tube, produced no evacuation of the bowels, but disturbed sleep, due to distress in the stomach and general exhaustion. To collapse, probably owing to repeated hemorrhages in the stomach, the patient finally succumbed.

XXX. P. K., æt. 35. Gastritis mucosa et gastectasia majoris gradus. Obstipatio. No passage for days.

Experiment 56.—Last passage 4 days ago. Doses up to 20 grm. Carlsbad Sprudel Salt produce no evacuation of the bowels.

Experiment 57.—Last passage 5 days before. A dose of 30 grm. Carlsbad Sprudel Salt in 250 c.c. distilled water, administered by the stomach tube, produced no stool, but also no pain. The bowels had to be forced open by means of infusion of senna.

XXXI. K. J., æt. 30, driver. Catarrhal disease of stomach, with atony of the large intestines. No passage for 3 or 4 days, causing much discomfort.

Experiment 58.—Last passage 3 days before. Five grm. Carlsbad Sprudel Salt were taken in 250 c.c. spring water at 18° C. After 9 hours a passage without any distress.

Experiment 59.—Last passage 3 days ago. Early, fasting, 5 grm. Carlsbad Sprudel Salt were taken in 250 c.c. Mühlbrunnenwasser at 50° C. No passage during the whole day, but peculiar sensations in the abdomen.

Experiment 60.—Last passage 2 hours before the taking of the salt. Early, fasting, 5 grm. Carlsbad Sprudel Salt were taken in 250 c.c. Mühlbrunnenwasser at 50° C. The same dose repeated in 15 minutes. The patient suppressed the desire to go to stool immediately after the second dose, therefore he had no passage until 1½ hours, and after 12 hours a second. Patient feels much relieved in the abdomen.

Experiment 61.—Last stool 3 hours before the trial. The same dose as above, substituting distilled water as the solvent. After the second dose desire to go to stool, which was suppressed by the patient; he, therefore, had no passage that day, but felt a dull sense of fullness in the abdomen during the day.

XXXII. H. Z., æt. 18. Obstipatio minoris gradus. No passage for 2 to 3 days; in addition to this, indefinite sensations in the abdomen.

Experiment 62.—Last passage the day before. Early, fasting, 5 grm. Carlsbad Sprudel Salt were taken in 250 c.c. Mühlbrunnenwasser. The patient remaining seated, after 1 hour a movement of the bowels, followed in 3 hours by a second pappy evacuation.

Experiment 63.—Last passage the day before. The same dose as above, with the exception that the patient took a walk. A desire to go to stool was felt in 15 minutes; this was suppressed by the patient, who, however, had a passage 1 hour later, and after four hours a second.

XXXIII. H. J., æt. 32. With the exception of an occasional sense of oppression felt in the chest, nothing abnormal could be detected. Bowels moved regularly once or twice daily.

Experiment 64.—Last passage 2 hours before taking the salt. Five grm. Carlsbad Sprudel Salt, dissolved in 250 c.c. Mühlbrunnenwasser, at 50° C., were taken early, fasting. The patient remained quietly seated. After

2 hours a passage, and during the day 2 more pappy evacuations. Feels weak and exhausted.

Experiment 65.—Last passage the day before. The same dose as above, with the exception that the patient takes a walk. After 3 hours a passage, in the afternoon a second, with weakness and lassitude.

Experiment 66.—Last passage 2 hours before the trial. The same directions as above, with the exception that distilled water was used as solvent. The patient promenades in the garden. After 4 hours a passage; in the afternoon a second pappy evacuation. The patient complained of slight headache.

XXXIV. R. P., æt. 37. Excepting nervous eructations, nothing abnormal to be detected. Bowels move daily, as a rule, but the passages are hard and voided with difficulty.

Experiment 67.—Last passage the day before. Early, fasting, 5 grm. Carlsbad Sprudel Salt were taken in 250 c.c. Mühlbrunnenwasser at 50° C. The same dose repeated after 15 minutes. The patient remained quietly seated indoors. Bowels moved in half an hour, followed during the day by a number of watery stools.

Experiment 68.—Last passage the day before. The same directions as above, with the exception that the patient takes a walk during the time of observation. After 15 minutes desire to go to stool, and after 30 the first evacuation of the bowels, followed during the day by three watery passages.

A number of further experiments produced the same result. But on continuing the trials the patient complained of a general sense of weakness and distress in the head.

SECTION 3.

GROUPING OF THE RESULTS FROM EXPERIMENTAL SERIES I.

1. Carlsbad Sprudel Salt must positively be acknowledged to be a remedy which expedites the movements of the bowels, for in 34 (really 40) cases examined it was only in 3 (XXVIII, XXIX, XXX) in which even maximum doses failed, and these all were severe cases of dilatation of the stomach.

2. The consistence of the stools is generally such that the first one is solid, or consisting, generally, of balls or masses of feces, the succeeding ones pappy to watery. The later stools are light yellow in color and not feculent, but smelling rather of sulphuretted hydrogen.

3. The time of the appearance of the first passage after taking the salt differs widely; in case XXVI, for instance, the first passage appeared in 15 minutes, in case XXXIV after half an hour, but in other cases later—in case XXXI even as late as 9 hours, when it was almost impossible to positively decide if the passage would have resulted even without the action of the

salt. The time of appearance of the stools was dependent upon a number of circumstances.

(*a*) Upon the nervosity or torpidity of the individual. In nervous individuals with a tendency to diarrhœa (III, XXVI, XXXIV), the passages may take place soon after the taking of the salt (after $\frac{1}{4}$–1 hour); in torpid individuals with atonic intestinal muscular development, however, much later, up to several hours (IX). The stools were also late in cases of catarrhal affection of the bowels. In cases of intestinal obstruction (pyloric or intestinal stenosis) the passages set in even later (in case XXIII as late as 9 hours), or even not at all (XXIX).

(*b*). The deportment of the patient after taking the salt influences the evacuations very much. The patients feel soon after taking the salt a desire to go to stool, and if they do not at once give way to it the evacuation will be delayed, as in cases XXXII, XXXIV, or may even fail to set in at all (XXXI, 61), when the accumulated fecal masses may either be demonstrated in the cæcum or in the left flexure.

(*c*) The quantity of the salt taken has, however, the greatest influence on the passages. This will be treated of in detail below.

4. The evacuations after taking the Sprudel Salt were, in the majority of cases, excepting a painless rumbling in the abdomen, generally painless and not followed by disagreeable symptoms. For of the 40 individuals observed, 28 bore the administration very well; in the other 12 cases the following was noticed :—

In cases of general nervousness (III, VI, XIV, XXVI) the patient complained of nausea, pain in the abdomen, and felt weak after the evacuations. In cases of partial intestinal impermeability (IV, XXII, XXIII, XXIX), as also in cases complicated with great dilatation (IV, XXIII, XXVIII, XXIX), severe abdominal complaints, as nausea, flatulency, eructations, even vomiting, sharp pain in the abdomen, and sleeplessness were produced. In one case (XXI) of acute catarrh of the bowels, taking of the salt was followed by burning, and passage of blood, as also in case (XV) of hæmorrhoidal knots. In case (XI) of mitral insufficiency, retching, oppression and palpitation followed the administration of the salt, so that its use had to be discontinued.

5. In reference to the dose, the following has been demonstrated :—

(*a*) The smallest quantity of Carlsbad Sprudel Salt tried was 5 grm., which the physician in private practice may assume to be one teaspoonful. This quantity of salt was given dissolved in 250 c.c. of water, and in this concentration is of a moderately salty, not unpleasant taste. In 27 individuals who took this dose fasting it only acted in 9 cases; seven of these cases (III, XII, XIX, XX, XXIV, XXVI, XXXII) were extremely nervous individuals. Of the 9 individuals cited, 2 (III, XXIV) had regular daily passages; in 3 cases (XII, XXVI, XXXIII) the passages were irregular, with a tendency to diarrhœa, and in 4 cases (XIX, XX, XXII, XXXIII) the bowels were constipated. The dose of 5 grm. generally only acted after the lapse of some time; only in two cases (XXII, XXVI) the bowels were opened inside of an hour, while in other cases it took 2–4, or even 6 hours. (XII). The dose of 5 grm. produced in 4 cases only one stool, whilst in 5 cases with a tendency to diarrhœa, 2–3 evacuations during the day.

(*b*) The dose of 10 grm. Carlsbad Sprudel Salt was tried, dissolved in 250 c. c. of water. This concentration, however, was not deemed practical, as the patients, on account of the bad taste and the nausea which generally followed, did not like to take it, so that the solution had to be given with the stomach tube (III). The solution of 10 grm. Carlsbad Sprudel Salt in 500 c. c. of distilled water was then tried, but the patient could not take such a large quantity of salty liquid at one dose; it caused nausea and retching (XIII, 29). It was, therefore, deemed expedient to administer the 10 grm. in two doses, of 5 grm. each, dissolved in 250 c. c. of water and given (fasting) half an hour apart. In 33 cases where this was tried it proved efficacious in 23, but in 10 cases the dose of 10 grm. remained without effect. Of these, 2 cases (IX, 18; XIX, 33) were cases of atonic, habitual constipation with idiosyncrasies against saline drugs; 4 cases of constipation combined with nervous ailments of the stomach, or gastric catarrh (V, 11; VI, 13; XVI, 36; XVIII, 37) and 4 cases of constipation connected with great dilatation of the stomach (VIII, 15; XXIII, 54; XXIX, 55; XXX, 56). The earliest at which the evacuations took place, ¼ to ¾ of an hour, after the administration of the first dose (X, 21; XXVI, XXXIV, 67, 68) the latest after 3 hours in case XVIII, 40, which was combined with gastric catarrh. The average time at which the bowels were moved was 1 to 2 hours after the taking of the salt; in 23 successful cases, 11 times. There were mostly a number of passages at different intervals when 10 grm. of the Carlsbad Sprudel Salt was taken. One passage only was produced in 3 out of 23 people; in these 3 cases (I, II, IV) the bowels were moved regularly every day. A greater number of passages was produced in 3 nervous individuals, having a tendency to diarrhœa (III, XXVI, XXXIV); in these cases 4 passages during the day followed the taking of the salt, as in these individuals the dose of 5 grm. already acted severely.

The accompanying symptoms attending the administration of 10 grm. Sprudel Salt, which quantity in private practice is equivalent to two teaspoonfuls, were observed in a few individuals. The above mentioned 3 nervous persons complained, after abundant passages, of lassitude, a general sense of weakness and headache; in case XXII, of partial impermeability of the bowel, vomiting and violent pain in the abdomen set in before the stools.

Two divided doses of 5 grm. each, and one 10 grm. dose of Sprudel Salt, which were tried alternately in 5 persons (III, 78; XIII, 29; XXXIV, XXXVI, XXXVII), did not alter the time of the appearance of the stools. There was greater frequency of the passages and longer continued action, therefore more complete cleaning out of the bowels, after divided than after one large dose.

(*c*) The dose of 15 grm. Carlsbad Sprudel Salt was either given at once in a solution of 250 c. c. of water, and in this case given by the stomach tube, or divided in 3 parts of 5 grm. each in 250 c. c. of water and given, fasting, at intervals of half an hour. This dose had no effect in 4 individuals out of 9 cases tried, in which smaller doses were without effect. Of these cases 3 were cases of great dilatation of the stomach (XXVII, XXVIII, XXIX), and 1 case (IX, 18) of severe habitual constipation which, however, were cured after one year. The action of doses of 15 grm. Carlsbad Sprudel Salt, in those cases in which they were attended with success, did not differ in the time at which they began to operate from the cases in which 10 grm. doses were given, but in the greater frequency and duration of the watery evacuations during the rest of the day. The former dose (15 grm.) was more frequently attended with disagreeable symptoms, such as vomiting, pain and burning in the abdomen. These symptoms were especially severe when the passages were delayed by an obstruction to the progress of the contents of the bowels (XXIII, partial impermeability), or the Carlsbad Sprudel Salt failed to act and accumulated in the intestinal canal. The dose of 15 grm. (equal to 3 teaspoonfuls) must, therefore, be considered the maximum dose for the therapeutic use of the Sprudel Salt.

(*d*) Larger doses, 20 to 30 grm. were also tried. In case IX, 19, of habitual constipation, 20 grm. were required to produce the desired effect without any severe accompanying symptoms. On the other hand, in cases of great dilatation of the stomach (XXVIII, XXIX, XXX) these or even larger doses failed. The salt produced in the first 2 cases only severe pain in the stomach.

6. The nature of the solvent used also influenced the effect produced by the Sprudel Salt.

(*a*) The different temperatures of the solutions (in cases II and III) showed no decided influence upon the result. A warmed solution of Carlsbad Sprudel Salt only had the disagreeable addition of being less agreeable to the taste and more apt to produce nausea and retching.

(*b*) Soda or Seltzer water used as solvent for the Sprudel Salt (in cases XIII, 27, 28; XIX, 41; XXI, 41) showed that, in comparison with ordinary water, the effect was much more prompt and continuous when the former were used than when ordinary water was employed, in addition Soda and Seltzer water made the solution much more palatable, so that much more concentrated solutions can be borne by the patients; I therefore, in private practice, order the Sprudel Salt dissolved in a carbonated water.

(*c*) To promote the purgative effect of the thermal waters at Carlsbad, Sprudel Salt is frequently added (5 grm. to a beaker). I have in 4 cases (XXV, XXXI, XXXIII, XXXIV) made trial of the Sprudel Salt in this way, using at one time 5 to 10 grm. Sprudel Salt in 250 c.c. distilled water at 50° C., and the next time the same quantity dissolved in 250 c.c. bottled

Mühlbrunnenwasser warmed to 50° C. The results were, however, even in the same individual, not always the same. At one time the solution of Sprudel Salt in Mühlbrunnenwasser was more efficacious; at another time the solution in distilled water (XXV); so that the question in this direction is not as yet settled. One thing, however, is certain, that 5 grm. Sprudel Salt act more energetically upon the functions of the intestines than a glass of Carlsbad Thermal water.

7. The influence of exercise on the action of the salt was studied during the summer in 3 cases, the patients on one day being required to remain quietly in their room and on the next to take a promenade in the clinical garden. The patients reported that during exercise, after taking the salt, they would feel a much more decided inclination to go to stool than when confined to their room, provided they would not repress this desire, when the evacuation would be delayed. It is, therefore, well to uphold the opinion that by exercise the action is expedited.

8. In the general condition of the patients, after the use of the Sprudel Salt, very little change was noticed in the majority of cases. The patients felt relieved after the evacuations; if the salt, however, was given for a number of days consecutively, then the patients complained of a sense of weakness, languor, confusion of ideas and headache, and became more anæmic (XXXIII, XXXIV).

SECTION 4.

SPECIAL EXPERIMENTS MADE TO DETERMINE THE NATURE OF THE EVACUATIONS UNDER THE INFLUENCE OF THE SPRUDEL SALT.

To gain a better insight into the *modus* of the appearance of the passages, I have made 5 examinations upon the 3 following individuals (which are to be added to Series I). The examinations were made of fractionally-collected passages. Inasmuch as I had determined by previous experiment that the filtrate of the normal fecal masses made with distilled water reacted alkaline, contained chlorides and traces of sulphates, making it impossible to detect the constituents of the Sprudel Salt that may have gotten into the passages, I have added to the Sprudel Salt in all the experiments 0.5 grm. potassium ferrocyanide or carmine. If the passages after the Sprudel Salt thus prepared were red or colored blue with ferric chloride, the conclusion must be drawn that the constituents of the Sprudel Salt must also be present.

XXXV. Sz. J., æt. 23, laborer, suffering from catarrhal gastritis. Bowels moved regularly every day, and contained no mucus.

Experiment 69.—Five grm. Sprudel Salt with 0.5 grm. ferrocyanide of potassium, dissolved in 250 c.c. of cold distilled water, were given early, fasting; in half an hour the identical dose was repeated. Within an hour rumbling in the intestines and inclination to go to stool, passage of intestinal gases; it was, however, not until $3\frac{1}{2}$ hours that a sausage-like stool was voided, which was not colored blue with ferric chloride. In 15 minutes there was another passage, of a mushy consistence and light-yellow color, which was rendered blue by ferric chloride. The patient had another passage on the following day, which was also colored blue by the reagent; and another passage on the third day, which gave a faint reaction only. There was no reaction on the fourth day. The following experiment was not made until the fifth day.

Experiment 70.—Early, fasting, 10 grm. Sprudel Salt and 0.5 grm. potassium ferrocyanide were administered, dissolved in 250 c. c. cold distilled water; in half an hour another 5 grm. Sprudel Salt in $\frac{1}{4}$ litre water, but without the addition of ferrocyanide of potassium, were administered. Soon after the first dose great rumblings, passing of wind and a desire to go to stool set in, whereupon, one hour after taking the first dose, a sausage-like stool, surrounded by a yellow, mushy mass, was passed. The sausage-like stool gave no reaction with ferric chloride, whilst the mushy mass gave a distinct blue color. During the day there were a number of watery mucous stools of light-yellow color, emitting the odor of sulphuretted hydrogen, in all of which the presence of potassium ferrocyanide could be demonstrated.

XXXVI. J. D., æt. 34, servant. Passages hard, generally every other day; otherwise healthy.

Experiment 71.—Early, fasting, 5 grm. Sprudel Salt in $\frac{1}{4}$ litre distilled water without the addition of potassium ferrocyanide were administered; after half an hour the same dose, but with 0.5 grm. of potassium ferrocyanide. An hour after taking the first dose a hard stool was passed, preceded by a slight rumbling and passage of wind. The matter passed had an alkaline reaction and the ferric chloride failed to produce the blue color. An hour thereafter a mushy stool was passed without giving the blue color. A third dose of 5 grm. Sprudel Salt with 0.5 grm. ferrocyanide of potassium was given. After half an hour there was a copious, watery, yellowish-green, bad-smelling evacuation, which contained pieces of undigested meat and which gave a strong blue color with the salt of iron. After $2\frac{3}{4}$ and $3\frac{1}{4}$ hours there were several more discolored, watery stools mixed with mucus, and having an alkaline reaction, in which the abundant presence of ferrocyanide of potassium could be easily demonstrated. The filtrate of the watery stools did not effervesce with HCl, but gave an intense cloudiness with $BaCl_2$.

Experiment 72.—On the following day the same patient took early, fasting, 10 grm. Sprudel Salt in $\frac{1}{4}$ litre distilled water without the addition of ferrocyanide of potassium. After $1\frac{1}{2}$ hours hard fecal masses were passed, and in 15 minutes thereafter a mushy, yellow, gelatinous mass. The solid gave with ferric chloride a distinct blue color; the mushy mass, however, barely gave a faint greenish blue coloration. After two hours there were two more watery passages, reacting alkaline, which gave no reaction whatever with the ferric salt; the filtrate, however, gave a decided precipitate with $BaCl_2$.

XXXVII. D. F., æt. 18, shoemaker. Perfectly healthy; bowels regular.

Experiment 73.—10 grm. Sprudel Salt dissolved in 500 c. c. of distilled water, the solution highly colored with carmine, were introduced into the stomach by means of the stomach tube. After a few minutes there were slight nausea and eructations, then rumblings and passage of flatus. After an hour a piece was passed having the form of a sausage and consisting of a thick, yellow mass, reacting alkaline. After half an hour there was a second almost watery stool, of a red color, which did not effervesce with HCl, but produced a decided cloudiness with $BaCl_2$.

The reaction of the urine was determined in many cases after the administration of the Sprudel Salt. The same invariably reacted alkaline several hours after the use of large doses. The urine was specially examined in the last three individuals, inasmuch as the same had to be collected carefully apart from the stools. In experiment 69, the urine still reacted acid after 4 hours, after 6 hours alkaline, and gave, acidulated with nitric acid a blue coloration with ferric chloride. Still, on the following day, potassium ferrocyanide could be demonstrated in the acid urine. In experiment 70 the urine after 2 hours reacted acid, after 3 hours neutral, and only after 4 hours alkaline, with a per cent. of ferrocyanide of potassium, and even on the day following traces of ferrocyanide could be demonstrated in the acid urine. In experiments 71 and 72, where there was no disease of the stomach, the urine reacted neutral after $1\frac{1}{4}$ hours, after $1\frac{1}{2}$ alkaline, and effervesced on the addition of HCl. This continued during 9 full hours after taking the salt. In the healthy individual, in experiment 73, the urine reacted alkaline in one hour after taking the Sprudel Salt. In passing, it may be here mentioned that in no case could the presence of ferrocyanide of potassium be detected in the saliva.

SECTION 5.

RESUMÉ OF THE EFFECT ON THE EVACUATION OF THE BOWELS AFTER THE ADMINISTRATION OF SPRUDEL SALT.

The total result of the researches on the above-mentioned 37 individuals, in reference to the appearance of the stools after taking Sprudel Salt, is the following:—

1. The action of the Sprudel Salt manifests itself soon after taking the same by rumblings in the abdomen, sometimes nausea, desire to go to stool, which must be heeded by the patient lest the passages be thereby delayed or even prevented.

2. Mostly after ¼–½ an hour there is passing of flatus, and frequently also eructations.

3. With increased desire to go to stool there follows, occasionally, in half an hour, a hard evacuation of fecal masses from the lowermost section of the intestine, followed, from the upper section, by mushy, and later on frequently by numerous watery passages, coming from the uppermost part of the intestinal canal.

4. In the first passage, generally, nothing abnormal can be observed; not until the second and the following can the constituents of the Sprudel Salt, bile and mucus be found. Of the constituents of the Sprudel Salt, the sodium sulphate and chloride can be detected, but not the bicarbonate.

5. The completeness with which the bowels are emptied depends, aside from individual peculiarities and the pathological condition of the intestinal canal, mainly upon the dose; 5 grm. may be considered the minimum one, 10 grm. the medium, and 15 grm. the maximum dose.

6. The bowels are generally evacuated without much trouble or effort.

7. Carbonated waters as solvents for the Sprudel Salt, as also exercise, facilitate the action.

The physiological significance of these facts will be specially treated in connection with the influence of the salt on the functions of the stomach. The clinical deductions will also receive a special chapter at the end of this treatise.

SECTION 6.

THE ACTION OF THE SOLUTION OF SPRUDEL SALT ADMINISTERED BY THE RECTUM.

The same was studied in three individuals.

XXXVIII. D. W., æt. 35, official. Suffering from habitual constipation; no passage for 2 or 3 days at a time.

(*a*) Ordinary lukewarm spring water was used in the experiments. The maximum quantity that could, by means of Hegar's funnel, be introduced into the intestine amounted to 2–2½ litres. These copious enemas were given on three evenings at bedtime. The patient could, when lying on the back, retain the liquid 5–10 minutes, after which water, with fecal masses, and at the end a yellow-colored, bad-smelling, mushy mass came away. A second passage followed during the night or on the following day, although the presence of liquid in the cæcum and in the flexures manifested itself by audible gurgling after the evacuation.

(*b*) A week after these preliminary experiments these abundant injections were continued during 2 weeks every second day, with the difference that a 1 per cent. solution of Sprudel Salt (10 grm. to 1 litre) was added thereto. The following differences manifested themselves in the use of the warm water and the solution of Sprudel Salt: At the first trial only 1 litre of the solution could be introduced, because the patient could not retain the same as well as the plain water. In the other trials the patient became accustomed to the same, so that 2 litres of the solution could be introduced, but it could not be retained longer than 5 minutes because of the desire to go to stool. The quantity of fluid fecal masses which came away after the use of the Sprudel Salt solution appeared to be much more abundant, and possessed a more penetrating odor than when plain spring water was used. The gurgling in the cæcum and flexures was very apparent. Following this, and preceded by rumblings, there was regularly, during the night, a second, mushy, yellow passage, having the odor of H_2S, and also on the following day there was another mushy evacuation. On the seventh and eighth days of the experiment the stools were mixed with streaks of blood and red-colored shreds of mucus, the patient feels a burning in the rectum during the evacuations, and the experiments were therefore discontinued.

XXXIX. H. S., æt. 54, hatter; well built and well nourished. No passage for several days. Injections of Sprudel Salt were used, not only on account of the habitual constipation, but rather on account of the *gall-stones*, which distended the gall-bladder so that it protruded 4 fingers below the ribs.

(*a*) The patient was directed to inject, every evening, 5 grm. Sprudel Salt in one-half litre of lukewarm water, and to retain the same as long as possible, even during the entire night. On the first evening the solution could not be retained, but passed out at once without any stool; it was, however, retained by the patient on the second evening, and was followed during the night by 2 passages, a hard and a soft one. On the third and fourth evenings it was followed during the night by a small evacuation; on the fifth day it was not followed by a passage, so that the patient had retained the solution the whole night.

(*b*) Ten grm. Sprudel Salt in one half litre lukewarm water were now ordered to be injected and to be retained as long as possible. The 5 days of observation showed the following: On the first evening the solution passed, with a mushy stool, after an hour; on the second there was a watery evacuation in an hour, and, not until morning, a second mushy one; on the third evening, after an hour, there was a watery passage; on the fourth a mushy one, after an hour, and another such in the morning. There were no traces of blood in the passages, but the patient complained of burning in the rectum. In this patient the urine of the night, examined in the morning, reacted alkaline and effervesced with *HCl*. The patient continued the Sprudel Salt injections (10 grm. to 1 glassful of spring water), and informed me that he went to sleep after the injection, and did not until morning have 2 mushy passages.

In passing, it may be mentioned that, owing to the continued use of the injections of Sprudel Salt, the long diameter of the gall bladder was reduced by 2½ finger-widths, and the sensitiveness of the swelling much diminished.

XL. E. M., æt. 25. Convalescent from perityphlitis. In consequence of the disease, an atonic weakness of the intestinal muscles had developed, so that there was no passage without some help. Warm spring water injections (1½ to 2 litres) were used; they, however, only caused but few fecal masses to pass with the stools; a sense of weight and fullness in the abdomen remained. One tablespoonful (20 grm.) Sprudel Salt was, therefore, ordered at one injection. After the solution of Sprudel Salt had been retained for a number of minutes, in addition to hard fecal balls, yellow fluid masses also came away; and during the night, or the next day, there would almost regularly be a thick, mushy evacuation. The patient felt herself invariably relieved after injections of Sprudel Salt, and always requested the addition of the same to the warm water. Later on, after the patient was able to leave her bed, the bowels regulated themselves.

SECTION 7.

RESUMÉ OF THE ACTION OF THE SPRUDEL SALT USED BY INJECTION PER ANUM.

The cited experiments prove that a solution of Sprudel Salt in the form of injection not only empties the lower bowel of its contents and washes away the mucus, but also exerts a powerful stimulation to the peristaltic action of the bowels. But, as after the first evacuation the solution still remained in the pockets of the intestines, it exerts a further stimulation upon the intestinal tube whereby an additional quantity of the contents of the bowel is passed downwards, and not only an evacuation is caused as in the case of the simple warm water, but is followed during the day by several mushy passages. When the solution of Sprudel Salt is retained a large part of the salt soon passes through the venous system of the pelvis into the urine, which is made alkaline, and the patient soon after the injection passes a quantity of clear, faintly acid urine. It is, therefore, best, when the pelvic organs are to be energetically acted upon, to give the Sprudel Salt by enema. Long continued use of injections of concentrated solutions of Sprudel Salt produces, in its train, additional disagreeable subjective and objective symptoms, as burning and severe straining, as also capillary hemorrhage from the mucous membrane.

I will treat of the effect of the long continued use of the Sprudel Salt *per os*, upon the functions of the intestines, farther on, when treating of the functions of the stomach.

II.—EXPERIMENTS MADE WITH CARLSBAD SPRUDEL SALT (POWDER FORM) ON THE FUNCTIONS OF THE STOMACH.

SECTION 8.

DESCRIPTION OF THE METHOD OF EXPERIMENTING.

A second series embracing 150 experiments were made upon 27 clinical cases to determine the influence exerted by the Sprudel Salt (powder form) upon the functions of the stomach. To determine this the following method was employed:—

The respective cases of the institution experimented upon were put upon the same diet and under observation for a number of days. The subjective as well as the objective symptoms on the part of the stomach and intestinal canal were, first of all, carefully observed and examined. To get as complete a picture of the condition of the stomach as possible, not only was the stomach capacity determined by means of the stomach volumeter of Jaworski, but also the chemical and mechanical function of the stomach were examined in a number of ways during the time of observation. In the first place the secretory power of the stomach was determined in a twofold manner by means of ice water. In the majority of cases 100 c. c. of distilled water brought to the freezing temperature by means of ice, and which I will, therefore, call ice water, was introduced into the stomach by means of the stomach tube, after the lapse of ten minutes 300 (in some cases 100) c. c. of distilled water at the temperature of the room were poured into the stomach by the same means. The contents of the stomach thus diluted, were then removed by means of Jaworski's stomach aspirator, measured and filtered. In the filtrate the free hydrochloric acid, mucus and digestive power and degree of acidity were determined. In other cases 200 c. c. of ice water prepared in the above manner were introduced into the stomach by means of a stomach tube attached to a funnel filled with ice, after ten minutes the gastric juice was aspirated without first diluting the same, and the contents of the stomach examined in the same manner as above. The details of the chemical examination I will speak of directly. On the following day the digestive chemism and mechanism of the stomach was tested by means of the egg albumen test: Early, fasting, the persons experimented upon took one hard-boiled egg without the yelk and 100 c. c. distilled water at the ordinary temperature. After 5 to 6 (sometimes after 2) quarters of an hour 100 c. c. of distilled water were poured into the stomach through the stomach tube, then aspirated and set aside for analysis, but the stomach was thoroughly cleansed by passing water through a rinsing apparatus so as to determine the amount of still undigested fragments of albumen present. The contents of the stomach first aspirated were measured, filtered, the amount of free hydrochloric acid and the degree of acidity of the filtrate were determined, and also tested for pepton, syntonin and mucus, in the manner described below.

The action of the solution of Sprudel Salt (powder form) upon the human gastric juice outside of the organism was observed. Specimens of 25 c.c. of an active gastric juice, obtained by means of ice water, were either neutralized with Sprudel Salt solution or brought to different degrees of alkalinity and agitated; hereupon it was acidified with hydrochloric acid, and brought to the acidity $= 1-10$ c.c. of one-tenth normal alkali per 100 gastric juice. The specimens in which the degree of alkalinity had been brought to above 4 c.c. one-tenth normal acid per 100 gastric juice failed, after being acidulated, to digest the albumen. The digestive ferment was therefore completely killed by the solution of Sprudel Salt, and its action could not again be restored by the addition of HCl. After having become acquainted with the digestive conditions in the stomach, with the number and condition of the stools of the individual under observation, preliminary experiments were made upon them.

To be able to determine the influence exerted upon the functions of the stomach, during the experiments with Sprudel Salt solution by the water introduced into the stomach, as well as the stomach tube, the experiments were first made, in the majority of individuals, with distilled water, observing the same conditions of quantity, temperature, time, etc., as in the case of the experiments made with solutions of Sprudel Salt (powder form).

The main experiments were made early in the morning, in the following manner: The patients were given, either to drink or by the stomach tube, solutions of 5, 10 or 15 grm. Sprudel Salt in 250 c.c. distilled, soda or Carlsbad Thermal waters early, fasting. Frequently larger doses of Sprudel Salt were given; 5 grm. were dissolved in 250 c.c. of water, and repeated at intervals of fifteen minutes. In both cases the patient either remained quietly seated or else walked about on the days of experimentation for 1–5 quarters of an hour, whereupon 100 c.c. distilled water, at the temperature of the room, were introduced into the stomach by means of the stomach tube, the contents of the stomach were then aspirated and placed into the receiving bottle. The liquid was measured, filtered, and the filtrate chemically analyzed in the following manner:—

For free hydrochloric acid: One drop of a 0.2 per cent. methyl violet solution was introduced into a beaker glass containing 10 c.c. of the filtrate. The violet coloration either shading into blue, or becoming decidedly blue to sky-blue, depending upon the amount of HCl present in the gastric fluid.

The tests of this reaction with gastric juice showed that a shading into blue was not produced until a per cent. of HCl was used which was equal to 3–4 c.c. of one-tenth normal alkali per 100 parts of liquid ; with 6 c.c. the coloration was almost blue, with 8 c.c. already intensely blue, and above that sky-blue. On the contrary, in samples of gastric juice which were acidified with lactic or acetic acid, no blue coloration could be observed by 3–4 times as great acidity. The reactions behaved in the same manner, also, when pepton or soluble albumen was present in the gastric juice. This reaction could not be used in those cases in which bile coloring matter had been mixed with the gastric juice. The acidity or alkalinity of the filtered gastric juice, as also of the solution of Sprudel Salt used, was determined with a one-tenth normal alkali, by boiling the solution with hydrochloric acid until a permanent red color was produced. Sulphates were tested for in the filtered liquid by means of hydrochloric acid and barium chloride, and the following grades or degrees observed : Precipitate, intense or slight cloudiness, intense or slight opalescence, traces. The solution of Sprudel Salt examined in this way showed a copious precipitate. To test for mucus, concentrated acetic acid was used, and the reaction classified as either cloudy or opalescent. The contents of the stomach were also tested for soluble egg albumen (syntonin) by means of officinal acetic acid and ferrocyanide of potassium, and the reaction classified as either cloudy or opalescent. The test for peptŏn, by means of a 5 per cent. solution of potassium hydrate and 1 per cent. solution of sulphate of copper, was performed in this way : To 10 c.c. of the filtered gastric liquid 2 c.c. potassium hydrate solution were added, then by means of a pipette so many drops of a solution of $CuSO_4$ were added until the liquid began to take on a rose-red coloring, which was the case when the gastric liquid contained traces of pepton ; after this there was added enough sulphate of copper solution until the red liquid began to take on a violet-blue shade. From the number of cubic centimetres of the solution of sulphate of copper needed the relative quantity of pepton could be determined.

The digestive power of the clear gastric fluid was tested in the following manner : In 2 small vials, each containing 25 c.c. of the gastric fluid, was placed a disk of egg albumen, 1 c.c. in diameter and 1 mm. in thickness, weighing 0.06 grm. (This was cut, by means of a double knife, from a recently hard-boiled hen's egg.) In one vial the liquid was acidified with 1 drop concentrated officinal hydrochloric acid, but not the other, and both placed in an air-bath of 38°–40° C. In case the gastric fluid reacted alkaline, it was first carefully neutralized with HCl, and then the drop of HCl was added. The time was now noted when the disk of albumen disappeared, and the liquid tested for pepton and soluble albumen after the digestion, in the manner described above. If the disk of albumen did not disappear after 24 hours or was dissolved with a putrid odor and no pepton reaction produced, the contents of the stomach were deemed completely devoid of digestive qualities, and, indeed, if the liquid acidified with hydrochloric acid did not digest, it was looked upon as being free from pepsin, and, in case the same was originally acid and still did not digest, it was decided that there was a lack of free HCl. Thereby it was also possible to decide if the acidity of the gastric contents was dependent upon free HCl or only upon organic acids. Direct experiments were made to determine the necessary degree of acidity of the gastric fluid for digestion. Repeatedly, 25 c.c. of acid gastric juice neutralized with one-tenth normal alkali and nearly neutral gastric juice were respectively brought to the acidity of 1, 2, 3 to 10 per cent. by volume one-tenth normal alkali per 100 of gastric fluid, and by means of the beforementioned disks of egg albumen the digestive power was tested. With an acidity of 1 and 2 per cent. by volume one-tenth normal alkali per 100 gastric liquid, no digestion was demonstrable, but a peculiar odor of fresh bread was observed. With an acidity of 3 there already appeared a slight pepton reaction, and the disk of albumen had for the most part disappeared after 24 hours. With an acidity of 4, the disk was dissolved in from 6 to 12 hours, and the pepton reaction was very marked. The higher the degree of acidity of the gastric fluid the more speedily did digestion progress, so that with an acidity of 10 the disk of egg albumen was dissolved in from $1\frac{1}{2}$ to 2 hours. If the contents of the stomach were originally strongly alkaline, it would occasionally dissolve albumen disks without first being acidified, but the digestive liquid contained no pepton and only dissolved albumen was demonstrable.

The chlorides were determined by testing the neutralized filtrate obtained from the contents of the stomach, to which 1 drop of chromate of potassium solution had previously been added (according to Mohr) by means of one-tenth normal silver nitrate solution. I have shown in the treatise, "Ueber das Verhalten des Kissinger und Karlsbader Wassers im Magen" (*Deutsch. Arch. f. kl. Med.*, B. xxxv), that this method of determination gives exact results, and that the substances containing N. do not prejudice the result, if the test be performed with the contents of the empty stomach. It is worthy of mention, however, that when the chlorides have been removed from the stomach they are not characterized by so sharp a color reaction as is the case in aqueous solutions of the same. An excess of silver nitrate is not followed by a red coloration, but the white color of the precipitate strikes a yellowish tint, which shows the point of saturation of the chloride with nitrate solution; when more is added the precipitate becomes orange, but not red. The c.c. of one-tenth normal silver nitrate solution to 100 c.c. of gastric fluid are given in trial Series II, column *j*, as chlorides.

The 26 cases examined in the same way are the following:—

Case	No. of Experiment	No. of grms. Sprudel Salt and cc. of Distilled Water Ingested	Temperature of the Sprudel Salt Solution in Degrees Celsius	No. of ½ hrs. the Solution was Allowed to Remain in the Stomach	No. of cc. of Gastric Fluid Aspirated after the Injection of 100 cc. Distilled Water	Condition of the Aspirated Gastric Liquid	Degree of Alkalinity	Degree of Acidity	Amt. of Chlorides Contained in	Test for: Free HCl with Methyl Violet	Sulphates with Barium Chloride	Mucus with Concentrated Acetic Acid	25 cc. of Gastric Fluid Digested the Albumen Disk: Without Addition of HCl	Pepton Reaction After the Digestion	After Addition of HCl	Pepton Reaction After the Digestion	Remarks
a.	b.	c.	d.	e.	f.	g.	h.	i.	j.	k.	l.	m.	n.	o.	p.	r.	s.
I.—L. L., æt. 20, soldier. No subjective gastric ailments	1	100 Distilled water.	Ice water.	10 min.	300 distilled water injected; 250 aspirated.	Transparent, pale wine yellow.	...	12.6	...	Very intense.	0	Barely trace.	After 5 hours.	Marked.	After 5 hours.	Marked.	The individual experimented upon was an Israelite.
	2	250 Distilled water. 5 grm. 250	18°	0	250	Colorless, faintly opalescent. Clear as water.	...	8.4	...	Decided trace.	0	0	"	"	After 2½ hours.	"	
	3	"	18° Distilled water.	2	96	"	...	6.4	...	"	0	0	"	"	"	"	
	4	"	"	2	110	Pale-wine Yellow. Clear as water.	...	7.2	...	"	Faint opalescence	0	After 16 hours.	Pale reddish.	After 12 hours.	Pale reddish.	
	5a	"	55° Distilled water.	2	105	"	...	3.6	...	Barely trace.	Faint cloudiness.	0	After 12 hours. Undigested after 24 hours	"	"	Indistinct	
	5b	250 Muehlbrunnen water.	55°	2	133	"	...	14.6	...	Very intense.	"	0	Entirely digested	Trace.	Entirely digested	Trace.	
	6	5 grm. 250	18° Soda water.	2	85	Yellowish, with mucus.	...	6.0	...	Trace.	Trace of opalescence.	0	Digestion incomplete.	Faint	After 12 hours.	Distinct.	
	7	5 grm. in 250; after ½ hour 5 grm. in 250.	18° Distilled water.	2+2	127	Without color: opalescent.	...	8.4	...	Distinct.	Opalescence.	0	"	Pretty distinct.	After 9 hours.	"	
	8	"	"	2+4	100	Colorless; slightly flocculent.	...	2.0	...	Not determinable	0	0	"	Trace.	After 5 hours.	"	

9								0				0	
II.—R. Z., æt. 30, merchant, Nervous dyspeptic gastric ailments	100 Distilled water.	Ice water.	10 min.	300 cc. distilled water injected; 400 aspirated.	Colorless; opalescent.	...	11.0	...	Very intense.	...		After 3 hours. Distinct.	After 2 hours. Distinct.	The individual experimented upon is an Israelite
10	250 Distilled water.	18°	0	225	Pale-wine yellow; opalescent.	...	7.6	...	Trace.	...	0	"	"	Before the aspiration no. 100 cc. of distilled water were not ingested.
11	"	18°	2	100	Clear as water.	...	3.6	...	Indistinct	...	0	After 12 hours. Faint.	After 3½ hours. "	
12	10 grm. 250	18° Distilled water.	4	113	Colorless; strongly opalescent; flocculent.	...	2.2	...	"	Decided cloudiness.	0	Undigested after 24 hours. 0	After 12 hours. Faint trace.	
13	"	55° Distilled water.	4	126	Colorless; strongly opalescent.	10.8	...	...	...	"	Trace.	Dissolved after 24 hours. 0	Undigested after 24 hours 0	
14	"	18° Soda water.	4	110	Colorless; clear.	...	0.8	...	...	Moderate cloudiness.	0	Undigested after 24 hours 0	After 12 hours. Pretty distinct.	
III.—K. R., æt. 26, clergyman. Chronic gastritis in the stage of hypersecretion of acid.	100 Distilled water.	Ice water.	10 min.	300 injected; 300 aspirated.	Lemon-yellow filtrate colorless.	...	10.8	...	Distinct.	...	Trace.	After 7 hours. Pretty distinct	After 4½ hours. Distinct.	
16	5 grm. 250	18° Distilled water.	2	180	Pale lemon-yellow; flocculent.	5.2	...	...	...	Decided cloudiness.	0	Undigested after 24 hours	Undigested after 24 hours Traces.	Odor of putrefaction.
17	"	55° Distilled water.	2	190	Lemon-yellow; transparent.	23.2	...	...	...	"	0	...	" 0	
18	5 grm. in 250 cc.; after ½ hour 5 grm. in	18° Distilled water.	2+4	85	Colorless.	...	1.2	...	Not to be determined.	0	0		After 6 hours. Faint.	
19	250 cc. 10 grm. 250	"	4	85	"	...	1.6	...	"	Slight cloudiness.	0	Undigested after 24 hours 0	After 7 hours. Barely a trace.	

37

Case.	No. of Experiment.	No. of grms. Sprudel Salt and cc. of Distilled Water Ingested.	Temperature of the Sprudel Salt Solution in Degrees Celsius.	No. of ½ hrs. the Solution was Allowed to Remain in the Stomach.	No. of cc. of Gastric Fluid Aspirated after the Injection of 100 cc. Distilled Water.	Condition of the Aspirated Gastric Liquid.	The Gastric Fluid — Degree of Alkalinity.	Degree of Acidity.	Amt. of Chlorides Contained in	Test for — Free HCl with Methyl Violet.	Sulphates with Barium Chloride.	Mucus with Concentrated Acetic Acid.	25 cc. of Gastric Fluid Digested the Albumen Disk. — Without Addition of HCl.	Pepton Reaction After the Digestion.	After Addition of HCl.	Pepton Reaction After the Digestion.	Remarks.
a.	b.	c.	d.	e.	f.	g.	h.	i.	j.	k.	l.	m.	n.	o.	p.	r.	s.
IV.—P. C., æt. 22, agriculturist. Chronic catarrh of stomach in the stage of hypersecretion of acid.	20	100 Distilled water.	Ice water.	10 min.	300 cc. Injected; 300 as aspirated. 272	Colorless, with yellow flakes.	...	10.0	...	Pretty distinct.	...	0	After 3 hours.	Distinct.	After 3 hours.	Distinct.	
	21	250 Distilled water.	18°	0	200	Lemon-yellow; yellow floccules.	...	4.0	...	Trace.	...	0	After 12 hours.	"	After 6 hours.	"	
	22	"	18°	2	250	Colorless.	...	12.8	...	Distinct.	Strong precipitate.	0	After 6 hours.	"	After 3 hours.	"	
	23	5 grm. 250	18° Distilled water.	0	190	Lemon-yellow cloudiness.	84.6	...	...	...	Cloudiness.	Trace.	Undigested after 24 hours	Faint.	Undigested after 24 hours	0	
	24	5 grm. in 250 cc.; after ½ hour 5 grm. in 250 cc.	"	2+4	200	Pale wine-yellow; opalescent.	9.2	6.8	...	Trace.	Precipitate.	Opalescence.	"	0	Incompletely digested after 24 hours	Pretty distinct.	The gastric liquid effervesces with HCl.
	25	100 Distilled water.	55° Distilled water.	10 min.	300 injected; 335 as aspirated.	Pale wine-yellow; transparent. Clear as water.	...	9.0	...	Distinct.	...	0	Digested.	Distinct.	Digested.	Distinct.	Bare trace
V.—P. S., æt. 32, soldier. Exacerbating acid bypersecretion, with general nervous phenomena.	26	100 Distilled water.	Ice water.	2	117	Colorless; opalescent.	...	13.0	...	"	Decided opalescence.	...	After 6 hours.	"	After 6 hours.	"	
	27a	5 grm. 250	18° Distilled water.	2	130	"	...	7.0	...	Trace.	"	...	After 20 hours.	Trace.	After 10 hours.	Trace.	Experiment made after a two weeks' treatment with Sprudel salt.
	27b	"	"														
	28a	"	55° Distilled water.	2	142	Colorless; opalescent.	...	1.6	...	...	Cloudiness.	...	Undigested after 24 hours	0	After 5 hours.	Faint.	

	28b		55°	2	110	Colorless; faintly opalescent.	...	13.2	Intense.	0	0	After 3½ hours. Distinct.
	29	250 Muehl-brunnen water. 5 grm. in 250 cc; after ½ hour 5 grm. in 250 cc.	20° Distilled water.	2+4	115	Colorless; containing mucus	...	13.6	Distinct.	Barely opalescent.	... Distinct.	After 3½ hours. "
VI.—S. S., æt. 22, shoemaker. Mechanical insufficiency, gastric with general nervous phenomena.	30	100 Distilled water.	0° Ice water.	10 min.	300 injected; 325 aspirated.	Colorless; opalescent.	...	6.6	Trace.	...	After 4½ hours. Pretty distinct	After 3½ hours. Pretty distinct
	31	250 Distilled water.	18°	2	150	Colorless; clear.	...	10.0	Distinct.	...	After 6 hours. Distinct.	After 4 hours. Distinct.
	32	5 grm. 250	18° Distilled water.	2	200	Faint wine-yellow.	...	6.4	Trace.	Milky cloudiness.	0 Undigested after 24 hours	0 Undigested after 24 hours
	33	"	18° Soda water.	2	155	Colorless; opalescent.	...	5.2	"	Faint cloudiness.	0 "	Barely a trace. Digested after 4 hours.
	34	"	18° Distilled water.	4	113	"	...	12.4	Distinct.	Faint opalescence	After 4½ hours. Distinct.	After 4½ hours. Distinct.
	35	"	"	6	140	"	...	13.2	"	0	After 3 hours. "	After 3 hours. "
	36	"	"	8	110	Colorless; clear.	...	5.8	Trace.	0	After 12 hours.	After 4½ hours. "
VII.—P. J., æt. 26, farmer. Without subjective gastric ailments.	37	100 Distilled water.	Ice water.	10 min.	300 cc. injected; 275 aspirated. 330	Pale wine-yellow; opalescent.	36.0	5.0	Barely a trace.	...	After 5 hours. Faint.	After 3 hours.
	38a	5 grm. 250	18° Distilled water.	2	114	Colorless; opalescent; mucoid.	1.6	...	...	Cloudiness.	"	Undigested after 24 hours
	38b	250 Muehl-brunnen water.	18°	2	85	Yellowish; clear.	2.0	...	...	"	"	After 12 hours. Faint trace.
	39	5 grm. in 250 cc; after ½ hour 5 grm. in 250.	18° Distilled water.	2+4		Colorless; faintly opalescent.		...	...	Opalescence.	Faint opalescence	After 16 hours. Faint.

39

Case	No. of Experiment	No. of grms. Sprudel Salt and cc. of Distilled Water Ingested.	Temperature of the Sprudel Salt Solution in Degrees Celsius.	No. of ½ hrs. the Solution was Allowed to Remain in the Stomach.	No. of cc. of Gastric Fluid Aspirated after the Injection of 100 cc. Distilled Water.	Condition of the Aspirated Gastric Liquid.	The Gastric Fluid			Test for			25 cc. of Gastric Fluid Digested the Albumen Disk.				Remarks
							Degree of Alkalinity.	Degree of Acidity.	Amt. of Chlorides Contained in.	Free HCl with Methyl Violet.	Sulphates with Barium Chloride.	Mucus with Concentrated Acetic Acid.	Without Addition of HCl.	Pepton Reaction After the Digestion.	After Addition of HCl.	Pepton Reaction After the Digestion.	
	a.	b.	c.	d.	e.	f.	g.	h.	i.	j.	k.	l.	m.	n.	o.	p.	r.	s.
VIII.—K.C., æt. 35, clergyman. Incipient gastric catarrh.	40	100 Distilled water.	Ice water.	10 min.	300 cc. injected; 400 aspirated.		...	5.0	...	Trace.	...	0	After 12 hours.	Faint.	After 3 hours.	Distinct.	
	41	10 grm. 250	18° Distilled water.	4	205	Pale wine-yellow; opalescent.	57.2	...	...	...	Precipitate.	Opalescence.	Dissolved after 24 hours.	Putrid odor.	Undigested after 24 hours	0	The gastric fluid effervesces with HCl.
	42	"	18° Soda water.	4	130	Lemon-yellow; mucoid.	24.8	...	...	...	Cloudiness.	"	"	"	"	0	"
	43	"	55° Distilled water.	4	210	Pale wine-yellow; transparent.	36.4	...	...	...	Copious precipitate.	"	"	"	"	0	
	44	"	18° Distilled water.	8	90	Colorless opalescent.	Neutral	...	...	...	Barely a trace.	0		Distinct.		...	
IX.—K.M., æt. 25, peasant girl. Hysteria with nervous gastric phenomena (vomitus nervosus).	45	100 Distilled water.	Ice water.	10 min.	300 cc. injected; 310 aspirated.	Colorless.	...	5.0	...	Trace.	...	0	After 5 hours.	Distinct.	After 9 hours.	0	
	46a	5 grm. 250	18° Distilled water.	2	170	Yellow.	3.1	...	...	...	Precipitate.	0	After 17 hours.		Undigested after 24 hours	Distinct.	
	46b	250 Muehlbrunnen water.	18° Distilled water.	2	120	Faint, yellowish; opalescent.	...	12.4	...	Very intense.	Barely a trace.	0	Undigested after 24 hours		After 5 hours.		
	47	5 grm. 250	18° Distilled water.	4	95	Wine-yellow; clear.	...	3.6	...	Barely a trace.	Faint opalescence	0		0	After 6 hours.		

No.	Description	Temp.	Time (hrs)	Vol.	Color/Appearance			Cloudiness				Remarks
48	15 grm. in 250; after ½ hour 5 grm. in 250.	"	2+2	101	Decided yellow color.	3.6	...	Trace.	Decided cloudiness.	0	"	0	Undigested after 24 hours	0	
49	"	"	2+4	110	Colorless; transparent.	...	5.6	Not determinable	Slightly opalescent. 0	0	After 7 hours.	Faint.	After 6 hours.	Faint.	
50	"	"	2+6	103	Colorless; slightly flocculent	...	2.4			0	Undigested after 24 hours	Trace.	"	Intense.	
51	100 Distilled water.	Ice water	10 min.	300 Injected; 320 cc. aspirated. 100	Colorless; faintly opalescent.	...	1.0	...	...	...	Undigested after 24 hours	...	After 12 hours.	Distinct.	
52	250 Distilled water.	14° Distilled water.	2		Colorless.	Neutral	...	...	...	...	"	Odor of putrefaction.	After 7½ hours.	Pretty distinct	After washing out stomach, Experiment 56 instituted.
53	5 grm. 250	"	1	263	Colorless; opalescent.	44.0	...	...	Precipitate.	0	Dissolved during 24 hours	0	Undigested after 24 hours	0	
54	"	"	2	102	Colorless.	Neutral	...	...	0	...	Undigested after 24 hours	...	After 12 hours.	Faint.	After washing out the stomach, Experiment 55 inmediately instituted.
55	"	"	2	247	Colorless; opalescent.	34.0	...	...	Precipitate.	...	Dissolved after 24 hours.	0	Undigested after 24 hours	0	The gastric fluid effervesces with HCl.
56	"	"	3	157	"	Neutral	...	...	0	...	Undigested after 24 hours	0	"	0	
57a	"	"	4	105	Colorless; mucoid.	2.0	...	...	0	Opalescence. Trace.	...	...	After 3½ hours.	Distinct.	
57b	250 Muehlbrunnen water.	20°	4	88	"	1.2	...	...	0		...	...	Incompletely digested after 24 hours.		
58	5 grm. in 250 cc.; after ½ hour 5 grm. in 250 cc.	14° Distilled water.	2+2	122	Colorless; flocculent	23.0	...	...	Cloudiness.	Opalescence.	...	...	Undigested after 48 hours	0	
59	"	"	2+4	93	"	9.2	...	...	"	"	...	...	"	0	
60	10 grm. 500	"	4	200	Colorless; decided opalescence	35.0	...	...	Precipitate.	"	...	...	"	0	

N.–J. W., æt. 32, official. Without subjective gastric ailments.

41

Case.	No. of Experiment.	No. of grms. Sprudel Salt and cc. of Distilled Water Ingested.	Temperature of the Sprudel Salt Solution in Degrees Celsius.	No. of ½ hrs. the Solution was Allowed to Remain in the Stomach.	No. of cc. of Gastric Fluid Aspirated after the Injection of 100 cc. Distilled Water.	Condition of the Aspirated Gastric Liquid.	The Gastric Fluid Degree of Alkalinity.	Degree of Acidity	Amt. of Chlorides Contained in	Test for Free HCl with Methyl Violet.	Sulphates with Barium Chloride.	Mucus with Concentrated Acetic Acid.	25 cc. of Gastric Fluid Digested the Albumen Disk. Without Addition of HCl.	Pepton Reaction after the Digestion.	After Addition of HCl.	Pepton Reaction After the Digestion.	Remarks.
XI.—W. W., æt. 30, farmer. Dilatation of stomach with remnants of food early in the morning	61	100 Distilled water.	Ice water.	10 min.	300 injected; 400 aspirated.	Colorless; mucoid.	...	0.2	...	...	...	Faint opalescence	Undigested after 24 hours	...	After 9 hours	Faint.	The stomach rinsed before the experiment.
	62	5 grm. 250	18° Distilled water.	4	115	Whitish; mucoid.	9.0	...	...	...	Cloudiness.	Opalescence.	"	0	Undigested after 48 hours	0	The stomach rinsed before the experiment.
XII.—M. C., æt. 24, farmer. Dilatation of stomach without subjective gastric ailments.	63	100 Distilled water.	Ice water.	10 min.	300 injected; 230 aspirated.	Colorless; opalescent.	Neutral	...	...	...	...	...	Undigested after 24 hours	0	After 7 hours.	Very faint.	
	64	5 grm. 250	18° Distilled water.	2	95	Whitish; cloudy; flocculent	7.2	...	...	...	Cloudiness.	Opalescence.	...	...	Undigested after 24 hours	0	
	65	"	"	4	75	Colorless; clear.	3.2	...	...	...	0	...	...	...	After 12 hours.	Faint.	
	66	5 grm. in 250; after ½ hour 5 grm. in 250 cc.	"	2+4	200	Colorless; mucoid.	3.2	...	...	...	0	Slight opalescence.	...	...	After 22 hours.	Very faint.	
XIII.—M. A. æt. 28, artist. Catarrhus ventriculi mucosus.	67	100 Distilled water.	Ice water.	10 min.	300 injected; 260 aspirated.	Colorless; opalescent.	0.2	...	...	...	...	Trace.	Undigested after 24 hours	Odor of putrefaction.	After 3 hours.	Distinct.	
	68	5 grm. 250	18° Distilled water.	2	75	"	1.2	...	...	...	Opalescence.	"	"	"	After 4 hours	Faint.	
	69	"	30° Distilled water.	2	215	Colorless; mucoid.	32.0	...	...	...	Precipitate.	"	"	"	Undigested after 24 hours	0	The gastric fluid effervesces with *HCl*.

70a	"	55° Distilled water.	2	85	"	4.0	...	...	"	Decided opales-cence. 0	After 12 hours.	Faint.
70b	250 Muehl-brunnen water.	55°	2	102	Colorless; opales-cent.	2.8	...	...	...		Undigest-ed.	Pretty distinct
71	5 grm. 250	18° Soda water.	2	70	Colorless; mucoid.	...	0.8	...	Opales-cence.	Opales-Trace. cence.	After 5 hours.	Distinct.
72	5 grm. in 250; after ½ hour 5 grm. in 250.	18° Distilled water.	2+4	80	Colorless; frothy.	...	0.4	...	"	0	Undigest-ed after 24 hours	Odor of putre-faction.	After 7 hours.	Faint.
73	100 Distilled water.	Ice water.	10 min.	150 injected; 350 as pi-rated.	Opalescent; mucoid.	2.5	...	...	"	"	Undigest-ed after 24 hours	Odor of putre-faction.	After 12 hours.	Distinct.
74	5 grm. 250	19° Distilled water.	4	Nothing In-jected; 210 aspirated.	Whitish; mucoid.	25.0	...	...	Precipi-tate.	"	Undigest-ed after 24 hours	...	
75	250 Distilled water.	19°	0	260	Brown; smeary food-rem-nants.	...	12.0	...	Only or-ganic acids.	...	Trace.	Undigest-ed after 24 hours	...	After 26 hours.	Faint.
76	5 grm. 260	18° Distilled water.	4	215	"	...	5.6	...	"	Precipi-tate.	Decided opales-cence.	"	0	Undigest-ed after 24 hours	Barely a trace.
77	10 grm. 250	"	12 hours	102	"	0.4	...	...	"	Opales-Opales-cence.cence.	...	...	"
78	100 Distilled water.	18°	0	80	Lemon-yel-low.	...	1.6	20.0	...	Trace.	Undigest-ed after 24 hours	0	After 4 hours.	Trace.
79	200 Distilled water.	Ice water.	10 min.	No water in-jected; 100 cc. as pi-rated.	Decided lemon-yel-low; clou-dy, filtrate yellow.	...	2.0	22.0	...	0		0	After 8 hours.	Distinct trace.
80	5 grm. 250	55° Distilled water.	4	120	Decided yel-low; filtrate yellow.	5.0	...	...	...	Cloudi-ness.	Distinct trace.	Undigest-ed after 24 hours	0	
81	5 grm. 250	10° Distilled water.	4	110	Faint lem-on-yellow.	1.0	...	12.5	...	0	Trace.		0	After 24 hours.	Faint.

XIV.—R. C., æt. 60, farmer. Dilatatio-ventricu-li majoris gradus. Food remnants fasting.

XV.—G. M., æt. 45, farmer. Carcino-ma and dilatation of stom-ach. Food remnants fasting.

XVI.—M L., æt. 40, single female with sub-jective gastric ailments and re-tarded di-gestion of albumen.

Notes: { The stomach completely freed from remnants of food with distilled water before the experi-ments. } { The 100 cc. dis-tilled water were not in-jected be-fore aspira-ting. }

43

CASE.	No. of Experiment.	No. of grms. Sprudel Salt and cc. of Distilled Water Ingested.	Temperature of the Sprudel Salt Solution in Degrees Celsius.	No. of hrs. the Solution was allowed to Remain in the Stomach.	No. of cc. of Gastric Fluid Aspirated after the Injection of 100 cc. Distilled Water.	Condition of the Aspirated Gastric Liquid.	Degree of Alkalinity.	Degree of Acidity	Amt. of Chlorides Contained in	Free HCl with Methyl Violet.	Sulphates with Barium Chloride.	Mucus with Concentrated Acetic Acid.	Without Addition of HCl.	Pepton Reaction after the Digestion.	After Addition of HCl.	Pepton Reaction after the Digestion.	REMARKS.
								The Gastric Fluid.			Test for		25 cc. of Gastric Fluid Digested the Albumen Disk.				
a.	b.	c.	d.	e.	f.	g.	h.	i.	j.	k.	l.	m.	n.	o.	p.	r.	s.
XVII.—H. A., æt. 33, Israelite. Healthy; digestion of albumen normal.	82	100 Distilled water.	20°	0	No water injected; 100 cc. aspirated	Colorless; faintly opalescent.	...	7.6	20.0	Distinct	...	0	After 5½ hours.	Pretty distinct.	...	...	...
	83	200 Distilled water.	Ice water.	10 min.	Nothing injected; 180 cc. aspirated	...	...	14.0	27.5	Intense.	...	0	After 2½ hours.	Distinct	...	...	...
	84	5 grm. 250	50° Distilled water.	2	120	Colorless; distinctly opalescent.	...	4.4	28.0	Faint.	Milky cloudiness.	Faint opalescence.	Incompletely digested after 24 hours.	...	After 3 hours.	Trace.	The patient has remained quietly seated during the time of observation.
	85	"	"	2	110	Colorless; faintly opalescent.	...	1.0	8.8	...	Decided opalescence.	0	Undigested after 24 hours.	0	After 12 hours.	Pretty distinctly	A walk in the garden during time of observation.
	86	250 Muehlbrunnen water.	50°	2	140	Colorless; distinctly opalescent.	...	3.0	37.3	Doubtful.	"	0	Incompletely digested after 24 hours.	Distinct trace.	After 7 hours.	0	Remained quietly seated during time of observation
	87	"	"	2	100	Colorless; opalescent.	...	3.6	11.0	Trace.	Moderate opalescence.	0	"	"	After 4½ hours.	0	A walk in the garden during time of observation
	88	5 grm. 250	50° Muehlbrunnen water.	2	160	"	19.2	...	33.0	...	Moderate cloudiness.	Trace.	...	...	Undigested after 24 hours	...	Remained quietly seated during time of observation.
	89	"	"	2	170	Faint lemon-yellow; opalescent.	22.0	...	33.0	...	"	Barely a trace.	...	...	...	...	A walk in the garden during the time of observation.
XVIII.—B. A., æt. 24, healthy Israelite. Com-	90	200 Distilled water.	Ice water.	10 min.	Nothing injected; 180 cc. aspirated 110	Colorless; strongly opalescent.	...	6.0	35.0	Trace.	Milky cloudiness.	0	After 6 hours.	Distinct	Undigested after 24 hours	...	...
	91	5 grm. in 250; after ¼ hour 5 grm. in 250.	50° Distilled water.	1+4	...	Colorless; frothy.	7.0	...	23.0	...	"	Trace.	...	...	...	...	A walk in the garden during the time of observation.

plains of no gastric ailments.	92	15 grm. in 50ºMuehl-brunnen water.	1+4	110	Colorless; 12.5 strongly opalescent	...	28.5	...	...	...	0	Remained quietly seated during time of observation.
		250; after ¼ hour 5 gr m. in 250.				12.5	21.0					
	93	"	1+4	120	"	...	...	...	"	"	"	A walk in the garden during time of observation.
	94	250 Muehl-brunnen water; after ½ hr. 250 Muehl-brunnen water.	1+4	140	Cloudy; faint yellow.	5.6	42.0	Trace.	Strongly opalescent.	"	Incompletely digested after 24 hours.	Trace.	After 6 hours.	Distinct traces	Remained quietly seated during time of observation.
	95	"	1+4	160	Faint yellowish-green.	5.0	38.0	Trace.	Milky cloudiness.	"	"	"	"	"	A walk in the garden during the time of observation.
XIX.— J. K., æt. 35, farmer; married. Moderate dilatation of the stomach with retarded digestion of albumen and acid hypersecretion.	96	100 Distilled water.	0	100	Colorless; faintly opalescent.	...	12.6	...	Intense.	...	0	After 4 hours.	Distinct trace.	...	
	97	200 Distilled water.	10 min.	Nothing injected; 180 aspirated.	Cloudy; lemon-yellow; flirme colorless.	...	19.6	42.5	Very intense.	...	0	After 3 hours.	"	...	
	98	250 Muehl-brunnen water.	4	130	Colorless; slightly frothy.	...	22.0	37.5	"	Decided opalescence.	0	After 2½ hours.	...	...	Remained quietly seated during the time of observation.
	99	5 grm. 250 50ºMuehl-brunnen water.	4	90	Transparent, shading into yellow.	...	3.4	13.5	Doubtful.	Very faint traces.	0	Undigested after 24 hours.	Faint traces	Digested after 3 hours.	Faint trace.	"
	100	"	6	90	Clear as water	...	2.0	7.5	"	Barely traces.	0	After 3 hours.	"	After 2½ hours.	"	"
	101	"	8	120	Colorless; faintly opalescent.	...	16.5	30.5	Very intense.	Distinct trace.	0	After 5½ hours.	Distinct	...	"	"
	102	50ºMuehl-brunnen water.	8	90	"	...	8.4	16.5	Distinct	Barely a trace.	0	After 7 hours.	Trace.	...	"	"
	103	250; after ¼ hour 5 grm. in 250. 50º	1+4	100	Clear as water	...	5.0	15.0	Not pronounced.	Trace.	0	Distinct trace.	After 6 hours.	Distinct trace.	"
	104	50ºMuehl-brunnen water.	1+4	620	Faint lemon-yellow; mucoid.	30.0	...	21.0	...	Precipitate.	0	Undigested after 24 hours	No odor of putrefaction. Trace.	Undigested after 24 hours	0	Experiment was made after two weeks' treatment with Sprudel Salt.
	105	100 Distilled water.	0	90	Colorless; transparent	...	7.2	2.0	Intense.	"	0	After 4½ hours.	Trace.	...	...	

Case	No. of Experiment	No. of grms. Sprudel Salt Water Ingested	Temperature of the Sprudel Salt Solution in Degrees Celsius	No. of ½ hrs. the Solution was Allowed to Remain in the Stomach	No. of cc. of Gastric Fluid Aspirated after the Injection of 100 cc. Distilled Water	Condition of the Aspirated Gastric Liquid	The Gastric Fluid			Test for			25 cc. of Gastric Fluid Digested the Albumen Disk				Remarks
a.	b.	c.	d.	e.	f.	g.	h. Degree of Alkalinity	i. Degree of Acidity	j. Amt. of Chlorides contained in	k. Free HCl with Methyl Violet	l. Sulphates with Barium Chloride	m. Mucus with Concentrated Acetic Acid	n. Without Addn of HCl	o. Pepton Reaction after the Digestion	p. After Addition of HCl	r. Pepton Reaction after the Digestion	s.
XX.—J. H., æt. 20, Israelite. Well nourished and well built. Digestion of albumen normal.	106	100 Distilled water.	18°	0	100	Colorless; faintly opalescent.	…	7.5	20.6	Distinct	…	0	After 3½ hours.	Trace.	…	…	Remained quietly seated during time of observation.
	107	200 Distilled water.	Ice water.	10 min.	100	Opalescent, shading into yellow.	…	15.0	36.0	Very intense.	…	0	After 2½ hours.	"	…	…	A walk in the garden during time of observation.
	108	250 Muehlbrunnen water.	50°	2	90	Clear as water	…	9.2	17.0	Distinct trace.	Milky cloudiness.	0	After 12 hours.	After 5 hours.	…	…	Remained quietly seated during time of observation.
	109	"	"	2	100	"	…	10.6	17.2	"	Opalescence.	0	"	Trace.	…	0	A walk in the garden during time of observation.
	110	5 grm. 250	50° Muehlbrunnen water.	2	130	Colorless; opalescent.	22.8	…	40.0	…	Milky cloudiness.	Trace.	…	…	Undigested after 48 hours.	0	
	111	"	"	2	100	Colorless; faintly opalescent.	5.2	…	30.0	…	…	"	Undigested after 24 hours.	…	"	0	
XXI.—R. Z., æt. 22, Israelite. Retarded digestion of albumen	112	100 Distilled water.	20°	0	140	Clear as water	…	2.0	…	Very intense.	…	0	After 1½ hours.	…	Digested.	Trace.	Remained quietly seated during time of observation.
	113	200 Distilled water.	Ice water.	10 min.	Nothing injected; 100 aspirated.	Colorless; strongly opalescent.	…	14.2	…	…	…	0	"	Distinct trace.			
	114	250 Artificial Sprudel Water.	14°	2	110	Colorless; faintly opalescent.	…	1.8	…	…	Strongly opalescent.	0	After 9 hours.	Trace.	…	…	Remained quietly seated during time of observation.

46

		14°	2	100	Clear as water	...	10.4	...	Intense.	Barely a trace.	0	After 4 hours	Distinct trace.	...	...	"
115	250 Natural Sprudel Water.															
XXII.—P. J., æt. 66, farmer. Digestion of albumen much delayed. Distress during digestion.																
116	200 Distilled water.	Ice water.	10 min.	Nothing injected; 170 aspirated.	Whitish; cloudy; flocculent.	...	18.4	43.0	Very intense.	...	0	After 3½ hours.	Distinct trace.	...	...	Remained quietly seated during time of observation.
117	250, After ¼ hour 250, after ¼ hour 250.	55° Sprudel water.	1+1+4	200	Colorless; strongly opalescent.	...	18.2	53.2	"	Milky cloudiness.	0	After 4½ hours.	"	...	...	"
118	"	"	1+1 +4	180	Whitish; cloudy; frothy.	...	20.0	51.0	"	Strongly opalescent.	0	After 3 hours.	"	...	...	
119	"	"	1+1 +4	150	Colorless; cloudy.	...	9.6	30.0	Distinct trace.	Distinct opalescence.	0	After 5 hours.	Distinct trace	...	...	A walk in the garden during the time of observation.
XXIII.—P. J., æt. 22, peasant girl. Catarrhal disease of stomach.																
120	100 Distilled water.	22°	0	90	Colorless; mucoid.	...	1.2	9.5	...	...	0	...	...	After 9 hours.	Not demonstrable.	
121	200 Distilled water.	Ice water.	10 min.	Nothing injected; 250 aspirated.	Colorless; strongly mucoid.	...	5.0	15.5	...	...	0	After 12 hours.	Pretty distinct.	...	...	
122	250 Sprudel water.	55°	2	150	Colorless; faintly opalescent.	11.5	...	11.5	...	Precipitate.	Trace.			Undigested after 24 hours	...	Remained quietly seated during time of observation.
123	"	55°	2	120	Whitish; cloudy; strongly mucoid.	...	0.8	18.0	...	Milky cloudiness.	Distinct trace.			Disintegrated after 24 hours.	0	A walk in the garden during time of observation.
XXIV.—M. T., æt. 14, Israelite. Digestion of albumen much delayed.																
124	100 Distilled water.	20°	0	100	Colorless; faintly opalescent.	...	7.5	17.5	Faint trace.	Precipitate.	0	After 4 hours.	Faint trace.	...	...	Remained quietly seated during time of observation.
125	250 Sprudel water.	55°	1	150	"	3.5	...	16.0	...	...	Trace.	...	...	After 24 hours.	...	
126	"	55°	1	200	Whitish; cloudy; flocculent.	...	5.2	33.0	Doubtful.	Milky cloudiness.	0	Undigested after 24 hours	Trace.	After 20 hours.	Distinct trace.	A walk in the garden during time of observation.

47

Case.	No. of Experiment.	No. of grms. Sprudel Salt and cc. of Distilled Water Ingested.	Temperature of the Sprudel Salt Solution in Degrees Celsius.	No. of ½ hrs. the Solution was Allowed to Remain in the Stomach.	No. of cc. of Gastric Fluid Aspirated after the Injection of 100 cc. Distilled Water.	Condition of the Aspirated Gastric Liquid.	Degree of Alkalinity.	Degree of Acidity.	Amt. of Chlorides contained in The Gastric Fluid.	Test for Free HCl with Methyl Violet.	Test for Sulphates with Barium Chloride.	Test for Mucus with Concentrated Acetic Acid.	25 cc. of Gastric Fluid Digested the Albumen Disk: Without Addition of HCl.	Pepton Reaction after the Digestion.	After Addition of HCl.	Pepton Reaction after the Digestion.	Remarks.
a.	b.	c.	d.	e.	f.	g.	h.	i.	j.	k.	l.	m.	n.	o.	p.	r.	s.
XXV.—L. B., æt. 27, israelite, goldsmith. Does not complain of any gastric pain.	127	300 Distilled water.	18°	0	350	Faintly opalescent.	...	11.0	...	Intense.		0	Digested.	Distinct.	Digested.	Distinct	
	128	100 Distilled water.	Ice water.	10 min.	300 injected; 310 aspirated.	Colorless; faintly opalescent.	...	18.0	...	"		0	After 5 hours.	"	"	"	Remained quietly seated during time of observation.
	129	250 Muehlbrunnen water. 5 grm in 250	53°	2	310	Faint bluish-green.	...	15.4	...	"	Faint cloudiness.	0	After 15 hours.	"	After 19 hours.	Distinct	"
	130	5 grm. in 250; after ½ hour 5 grm. in 250.	20° Distilled water.	2	370	Yellowish; transparent.	...	0.8	...	"	Milky cloudiness.	Trace.	Undigested after 24 hours	Odor of putrefaction.	Incompletely digested after 24 hours.	Faint trace.	"
	131		20° Distilled water.	2+4	115	Colorless; opalescent; frothy.	...	1.6	...	"	0	Barely a trace.	"	0	After 3 hours.	Trace.	
XXVI.—N. W., æt. 47, shoemaker. Acid catarrhal disease of stomach.	132	100 Distilled water.	19°	0	Nothing injected; 300 as aspirated.	Colorless; opalescent; flocculent.	...	13.6	...	Intense.	Pepton-syntonin reaction.	0	After 4 hours.	Faint trace.	After 4 hours.	Faint trace.	
	133	200 Distilled water.	Ice water.	10 min.	"	"	...	17.2	...	"	0	0	After 3 hours.	"	Digested.	"	
	134	Egg albumen and 100	20° Distilled water.	2	250	Colorless; strongly opalescent; large quantity of fragments of albumen.	...	21.2	...	"	Trace.	Strong opalescence.	After 5 hours.	Distinct	After 5 hours.	Distinct	Testing of the function of the stomach before treatment with Sprudel Salt.

No.	Test meal	Temp.	Time	Injected	Amount	Description		Value		Decided traces		After (1st)	After (2nd)	Notes
135	"	"	6		300	Pale yellowish; small quantity of fragments of albumen.	...	42.0	Very intense.	Decided traces.	"	After 4 hours.	After 4 hours.	
136	Meat soup, bread, beef steak. (*After Leube.*)		7 hrs.		200	Thickish; discolored meat fibres and pieces of bread.	...	40.2	"	"	Barely a trace.	After 1½ hours.	After 2 hours.	Examination of the gastric function during the treatment with Sprudel Salt.
137	100 Distilled water.	20°	0		130	Colorless; faintly opalescent.	...	4.6	Still traces.	0	0	Digested.	Distinct trace.	
138	Egg albumen and 100	20° Distilled water.	2		300	Yellowish; cloudy; large quantity of swollen fragments of albumen.	...	9.7	Faint trace.	Distinct trace.	Opalescence.	After 12 hours.	After 6 hours.	
139	"	"	6		275	Yellowish; cloudy; small quantity of fragments of albumen.	...	26.8	Intense.	Large quantity	Trace.	After 4½ hours.		Examination of the function of the stomach during the treatment with Sprudel Salt.
140	200 Distilled water.	Ice water.	10 min.	Nothing injected; 240 aspirated.		Colorless; opalescent; mucoid.	...	11.2	"	0	0	Digested.		
141	Meat soup, bread, beef steak. (*After Leube.*)		7 hrs.		150	Cloudy; few meat fibres, oil globules; odor of sebasic acid.	...	24.0	"	Distinct trace.	...	After 1½ hours.	Intense.	
142	100 Distilled water.	20°	0		165	Whitish; strongly opalescent; mucoid.	...	4.0	Barely a trace.	0	0	After 15 hours.	After 3 Distinct trace.	Testing of the gastric function after one month's treatment with Sprudel Salt.
143	200 Distilled water.	Ice water.	10 min.	Nothing injected; 220 aspirated.		Colorless; faintly opalescent.	...	3.0	Doubtful.	0	0	After 22 hours.	After 2 hours.	
144	Egg albumen and 100	20° Distilled water.	6		160	Colorless; very minute quantity of small pieces of albumen.	...	6.5	Trace.	Distinct trace.	Barely a trace.	Disintegrated in 24 hours.	After 3½ hours.	

SECTION 9.

DEDUCTIONS FROM THE TABLE OF EXPERIMENTS.

If the experimental results are examined, they will be found to differ widely. The influence of the salt on the functions of the stomach is therefore not always the same. It depends upon three circumstances: (*a*) above all upon the pathological condition of the stomach, (*b*) upon the dose of Sprudel Salt, and (*c*) upon the circumstances accompanying the ingestion of the salt, as solvent, and temperature of the same. In reference to point (*a*) this shows a difference of deportment in the Sprudel Salt, depending upon the presence of (α) an alkaline or slightly acid gastric secretion (X–XVI, also XXIII), (β) an acid gastric secretion (VI–IX, also XVII, XVIII, XX, XXI, XXIV), (γ) an excessively acid secretion (I–V, also XIX, XXII, XXV, XXVI), or lastly upon (δ) a high grade of gastric dilatation (XI, XII, XIV, XV, XIX).

1. When does the Sprudel Salt disappear from the stomach? is the question that obtrudes itself upon our notice in experiments of this kind. Placed in this wise it is, however, impossible to solve it, inasmuch as the different constituents of the Sprudel Salt do not disappear at the same time, because the resorptive power of the gastric mucous membrane differs for the different salts, as I have demonstrated is the case in man (*Zeitschr. f. Biol.*, XIX, p. 398). To obtain comparative results it is, therefore, necessary to confine ourselves to the disappearance of one constituent of the Sprudel Salt. This is not possible with sodium bicarbonate, as it undergoes a chemical change produced by the acidity of the stomach in addition to resorption. The sodium chloride, which could be easily determined quantitatively in the gastric juice, has the disadvantage that the chlorides already present in it materially influence the results obtained. There remains then only the third constituent, the sulphates, that could be used as a test for the disappearance of the Sprudel Salt. The sulphates, which are entirely foreign in the normal condition of the contents of the stomach, are not influenced by the gastric acid and can be very easily and distinctly detected by means of barium chloride.

(*a*) If, therefore, the disappearance of the sulphates from the stomach are taken as indicative of the disappearance of the Sprudel Salt, then column 1 shows that when 5 grm. doses were employed they disappeared in half an hour only in two cases (I, 3; X, 54), in which there was no disease of the stomach present; in all other cases, however, much later. In the majority of the cases the sulphates disappeared from the stomach entirely in the fourth quarter of an hour, *i. e.*, within an hour. Only in cases of dilatation of the stomach their disappearance is markedly delayed. After one hour large quantities of sulphates were still to be found in all cases of dilatation (XIV, 74; XV, 76; XI, 62), and in the case (XIX, 101) of dilatation of the stomach, which was examined after two hours, they were still present in small quantities. In all experiments where 10 grm. doses of Sprudel Salt were given the table shows that still after one hour a large quantity of Sprudel Salt was found in the stomach. In all cases the chloride of barium causes a cloudiness or precipitate in the gastric fluid. If the 10 grm., as was generally the case, were divided into two 5 grm. doses and given half an hour apart, there was much less sulphate found when the stomach was examined after the lapse of 2 or 4 quarter-hours after the administration of the last dose than when one large 10 grm. dose had been given (X, 58-60). The time of disappearance of the 10 grm. doses varied greatly. In case of acid gastric secretion (I, 8; V, 29; IX, 49; XII, 66) the sulphates had almost entirely disappeared after 6 quarter-hours. In other cases, as in acid catarrhal disease (VIII, 44) the sulphates disappear at the end of the second hour, and in dilatation of the stomach (XV, 77) sulphates may even be found on the following day. It may, therefore, be assumed that in the majority of cases a dose of 5 grm. Sprudel Salt disappears from the stomach in 4, and a 10 grm. dose after 6 quarter-hours. In some pathological conditions the disappearance may be very perceptibly delayed.

Inasmuch as larger quantities of Sprudel Salt required still longer time to disappear from the stomach, and as it is rather difficult to keep the patient fasting for hours, and as in addition to this, as will be observed below, larger quantities of Sprudel Salt very much affected the gastric function, I have forborne from making any experiments in this direction with larger doses than 10 grm.

(*b*) It is a rather delicate thing to speak of the disappearance of the chlorides from the stomach, as I have already mentioned. I have endeavored to determine the same for other purposes in some cases, and placed them in the table, column *j*. Only this much may be inferred therefrom, that, after the lapse of several quarter-hours after the ingestion of the solution of Sprudel Salt of the strength employed, 82.6 c. c. one-tenth normal silver solution to 100 salt solution, according to my method, the percentage of chlorides was usually found to be smaller than in the empty stomach, and invariably smaller than when obtained by the ice-water method (XVI, 78, 79, 81 ; XVII, 82, 83, 85 ; XVIII, 90, 91 ; XIX, 97, 101, 103; XXIV, 124, 125). The percentage of chlorides in the gastric contents after the taking of Sprudel Salt diminishes to a certain minimum (XIX, 101, 102), whereupon it begins to rise again.

At this place mention may be made of the circumstance, to which I now give more attention in my examination of the stomach, that the secretion of the chlorides from the gastric mucous membrane also increases in abnormal acid hypersecretion, and that in greater degree than the secretion of acid, whilst in acid insufficiency the quantity of chlorides also appears diminished. Compare, for instance, the cases XIX, 97 ; XXII, 116 ; XXIII, 120, 121.

(*c*) The disappearance from the stomach of the bicarbonates, the percentage of which corresponds to 90.4 c. c., one-tenth normal acid solution per 100 c. c. salt solution, is easily determined by the cessation of the alkaline reaction of the gastric contents, and does not go hand in hand with the disappearance of the sulphates, because, in addition to their absorption and mechanical removal from the stomach, they continually suffer chemical decomposition by the gastric juice. The length of time the bicarbonates remain in the stomach is, therefore, relative, according to the case. When 5 grm. or more of Sprudel Salt are given at one dose the carbonates may already disappear after the first quarter-hour if they happen in an acid gastric secretion, as I have repeatedly assured myself in cases of gastric dilatation with acid hypersecretion, associated with retention of food remnants. If these cases be eliminated then the table will show, column k, that in the nine cases cited suffering from acid hypersecretion the bicarbonates have already disappeared before the lapse of half an hour; in the four cases designated as moderate acid secretion and approaching nearer to the normal, this occurred in two (VI, 32; XVII, 84), in the other two cases (VII, 38; IX, 46) the gastric contents were at this time strongly alkaline, owing to sodium carbonate. In the stomachs with acid insufficiency, on the contrary, the sodium carbonate only disappeared in half an hour in the single case (X), and in every other case decided alkalinity is noticed, in the cases XIV, 74, and XVI, 80, even 26 and 5 degrees alkalinity.

When a 10 grm. dose is administered the carbonates will disappear in the following manner: In all cases of acid hypersecretion the carbonates have disappeared from the stomach after one hour. In the 4 cases examined of moderate acid secretion this was not the case in one instance; the acid contents even effervesced with HCl (VIII, 41). Not until five quarter-hours in case XVIII, 94, 95; after six quarter-hours in case VII, 39, and after two hours in case VIII 44, did the carbonates leave the stomach. In cases of acid insufficiency the time of disappearance was extended even more, but not so much as might be expected. For, as noticed in case XIII, 72, it was almost as long as in acid secretion. As an average, it may be stated that the carbonated alkalies of the Sprudel Salt will disappear from the stomach when 5 grm. doses are given in three quarter-hours, and when 10 grm. doses are given at the end of two hours. In all cases in which the carbonates were found larger quantities of sulphates were also present, so that it appears that the bicarbonates of the Sprudel Salt are more quickly absorbed in the intestinal canal than the sulphates, inasmuch as no carbonates could be found in the passages, but only sulphates, as is mentioned in Section 4, whilst the urine is found to be alkaline upon examination after the ingestion of Sprudel Salt, and the carbonates could be demonstrated by the effervescing of the urine upon the addition of an acid.

2. The *chemical function of the stomach* undergoes a decided change in every direction under the influence of the Sprudel Salt.

(*a*) The acidity of the gastric juice, as would be expected and may be seen in column i, is changed to alkalinity. This takes place only in the first moments, the alkalinity thereupon decreases, and the gastric acid becomes free. The time when free acid again appears depends upon the individual case and upon the dose employed. With a dose of 5 grm. Sprudel Salt the acidity of the gastric juice in strongly acid secretion, with the exception of case III, was very marked in half an hour after the administration of the salt. In case IV, 22, it was not only greater than that of the empty stomach, but exceeded even the acidity obtained by the egg albumen method. In cases of moderate acid gastric contents, on the contrary, the gastric acid was regenerated in two cases after half an hour, but not in two others, in which it required one hour (IX, 47). In alkaline gastric juices no strongly marked acid secretion follows the disappearance of the carbonated alkalies, still, in one case of alkaline secretion (XIII, 71, 72) a slight tendency to acid secretion was noticed.

When 10 grm. doses of Sprudel Salt are given the appearance of free gastric acid is much delayed. In strongly acid secretions the gastric acid is regenerated, on an average, in 1 hour after the ingestion of the salt, in moderately acid after 6 quarter-hours (IX, 49), or even later—in case VIII, 44, after 2 hours. In weak acid gastric secretion a decided alkalinity of the gastric juice is still present after 6 quarter-hours (X, 59). The regeneration of the gastric acid begins, therefore, under ordinary circumstances, after the administration of a 5-grm. dose in 3 quarter-hours, and after a 10 grm. dose at the end of 2 hours.

The secretion of acid under the action of the Sprudel Salt, however, shows certain peculiarities. In cases VI and IX, where the degree of acidity of the gastric juice was examined after 2, 4, 6 and 8 quarter-hours, it will be seen that the acidity rises to a certain height much stronger than when obtained by the ice-water method (experiments 30, 35), and falls from this maximum very quickly to even below the degree of acidity of the empty stomach (experiment 31, 36), so that, in reference to the action of the Sprudel Salt upon the secretion of acid, 3 stages may be observed: the stage of latent acidity, the stage of increase of acidity, and the stage of fall or reduction of acidity. This great cycle of secretion probably does not last much longer than 2-3 hours, depending upon the dose administered.

(*b*) The digestive power of the gastric juice, which is measured by the time required to digest the egg-albumen disk, shows, under the influence of the Sprudel Salt, very characteristic changes. An alkaline gastric juice, of course, does not digest at all of itself. But it is only exceptionally a pathological condition of the stomach in which the alkaline contents of the stomach would not digest an albumen disk after acidification with HCl, so that there are only exceptional cases in which the digestive ferment is totally absent (XIII, 67; XIV, 73). If the solution of Sprudel Salt, however, is put into the stomach, the gastric juice in all cases loses its digestive power so completely (as was found in trials outside of the organism, as cited in

Section 8) that it, even after decided acidification with HCl, failed to affect the albumen disk at all, as is demonstrated by the numerous trials on the table, column p. A gastric juice rendered strongly alkaline by means of solution of Sprudel Salt may of itself dissolve an albumen disk (II, 13; VIII, 41; X, 53, 55), but does not digest the same, as no pepton can be demonstrated in the solution. If the degree of alkalinity diminishes to a certain point after the imbibition of the Sprudel Salt, then it may, in rare cases (XII, 65, 66; XIII, 70), show a slight digestive power after acidification with HCl. This degree of alkalinity lies between 3 and 4; in other cases, however (IX, 46, 48), no trace of digestive power manifests itself at this degree of alkalinity. With the appearance of free gastric acid the digestive power is slowly regenerated. To insure a complete digestion of the albumen disk, however, the degree of acidity must at least reach to 5–6 (VI, 36; IX, 49). But even at this degree of acidity, at which the contents of the stomach obtained from the empty stomach or by means of the ice-water method, would invariably digest, in other cases the gastric juice called forth by means of the Sprudel Salt appears incapable of digestion (1, 2, 6, 7, IV, 21, 24; VI, 30, 32, 33; XVIII, 94, 95), and the acidity had to rise much higher or HCl had to be added to obtain a complete digestion of the albumen disk. It would therefore appear that either the gastric acid is more rapidly regenerated than the digestive ferment, which would point to free acid, and by the Sprudel Salt more strongly affected pepsin-forming anatomical elements, or that the salts still present in the stomach prevent the action of the pepsin, as may be seen by the columns n, o, in the experiments I, 7; IV, 24; VI, 32, 33; as long as the contents of the stomach show a greater percentage of salt, the gastric juice cannot be made thoroughly efficient, even when HCl is added (IV, 24; VI, 32; IX, 46, 48; XVII, 84). The condition found in the experiments, that the maximum of digestive power does not rise with the maximum of the acidity of the gastric juice, but rather generally falls with the same, will be hereby understood. A complete digestive power of the gastric juice after a dose of 5 grm. Sprudel Salt sets in, in cases of hyperacid secretion, with the exception of case III, after 2 quarter-hours; not then, however, in cases of moderate acid secretion. This, however, took place in the two experiments instituted for the purpose in case VI, 34, after one hour, but not in case IX, 47. Where the secretion reacted alkaline, naturally no gastric juice capable of digestion could be obtained in any case, not even, as in case XV, where the gastric contents reacted acid, which was, as was demonstrated in that case, not due to HCl, but depended upon organic fermentation acids.

After 10 grm. doses of Sprudel Salt the appearance of a gastric juice thoroughly capable of digesting was very much delayed. In cases of acid hypersecretion a gastric juice capable of digesting appeared after 5 quarter-hours only in the one case, XIX, 103, and after 6 quarter-hours in case V, 29; in other cases, as, for instance, IV, 25, it failed to appear even after 8 quarter-hours. Of course, this was still less the case in cases of moderate acidity, where, for instance, in case VIII, 44, of pretty decided acid reaction, the gastric juice was neutral after two hours.

. For the majority of cases it may, therefore, be assumed that a gastric juice capable of digesting reappears in the stomach after the administration of a 5-grm. dose of Sprudel Salt in one hour, and after a 10-grm. dose after two hours.

4. The gastric juice obtained under the influence of the distilled water and the Sprudel Salt solution shows still other differences which show themselves already in the appearance of the gastric contents.

(a) The gastric liquid obtained by means of distilled water appears, in the majority of cases (for instance, in cases II, XX, column g), more cloudy and full of suspended particles than when obtained after the ingestion of solution of Sprudel Salt. The solution of Sprudel Salt dissolves the morphological constituents (dead cells, fermentation organisms) from the walls of the stomach, and contains the same not suspended, but dissolved or in a semi-dissolved condition. The gastric fluid, therefore, appears the clearer after the ingestion of the Sprudel Salt, the longer time is allowed to elapse before it is removed from the stomach; *i. e.*, the longer the morphological constituents were in contact with the solution of Sprudel Salt.

(b) The Sprudel Salt generally produces a ropy, mucous, gastric fluid, which is difficult to filter; which, when alkaline or slightly acid, invariably shows traces of mucus (Table, column g), and that the more, the greater the quantity of Sprudel Salt has been introduced into the stomach (IV, 20, 24, 25; VII, 37, 38, 39; XVII, 82, 88, 89; XVIII, 90, 91, 92). With greater acidity of the gastric fluid, the mucus is, however, not demonstrable with concentrated acetic acid, although it certainly is present therein; for the presence of an albuminoid in the gastric juice is demonstrated by the violet-red reaction after addition of solution of sulphate of copper to the gastric fluid neutralized with potassium hydroxide. The mucus has probably undergone a certain modification, not yet fully determined, in the acid gastric contents. As a rule, no reaction for mucus is obtained in the filtrate of a gastric fluid obtained by distilled water, although the same may frequently be seen in the form of thread-like masses or mucous flakes—these remaining on the filter.

(c) The color of the gastric fluid (Table, column g) appears, under the influence of the Sprudel Salt, more frequently yellow than when distilled water only has been employed. Especially that aspirated in the first half-hour after the ingestion of the salt is colored yellow. This yellow color is caused by the pouring in of bile into the stomach, which points to the fact that a copious secretion of bile, and strong peristaltic movements of the upper portion of the intestinal canal, is produced by the action of the Sprudel Salt, which is really the case in the beginning of the action, owing to the stimulus of a warmer and more concentrated solution. If the bile has but recently been poured into the stomach, or perhaps been poured in during its excretion, then it is light yellow, and the filtrate is of the same color. But if the bile has been retained in the stomach for a longer time, then yellow or green flakes are formed. The gastric fluid appears, according to the acidity, either greenish-yellow or bluish-yellow, and the filtrate of the same color or else rosy-red.

(*d*) The digestive power of the gastric juice obtained by means of distilled water, especially ice water, shows itself much more active than when obtained by means of Sprudel Salt, when the samples are compared and are of the same degree of acidity, and that sufficient for digestion. This refers as well to the rapidity of peptonization as also to the reaction for pepton. The time for peptonization in the table, column *n*, I, 27; VI, 30, 32, in spite of the same degree of acidity of the gastric juices in both cases, was invariably longer when Sprudel Salt was used than when distilled water was employed, and this is also frequently the case when the acidity after Sprudel Salt exceeds that after distilled water, as in cases IV, 21, 24; XX, 106, 108. In the completeness of the power of peptonization a still greater difference was observed in the two kinds of gastric fluid. If the gastric juice obtained by means of Sprudel Salt solution without or after acidification completely dissolved the albumen disks, the digestive fluid failed to give a red color reaction of pepton, as was the case when obtained by distilled water, but a violet, or a shade of violet, as shown in the cases VI, 30, 33, and IX, 45, 49, column *n*, *o*, *p*, *r*. It is therefore very probable that the gastric juice obtained by means of Sprudel Salt cannot produce the final product of digestion pepton, but stops at the intermediate product of digestion, the propepton, which modification of albumen lies between the acid albumen and the pepton proper.

5. The action of cold and warm Sprudel Salt solution on the function of the stomach may be seen in the experiments I, 4, 5; II, 12, 13; III, 16, 17; IV, 24, 25; V, 27, 28; XIII, 68, 70; XVI, 80, 81, in which the Sprudel Salt was dissolved once in distilled water at 18°, another time at 55°. When a warm saline solution was administered, a much larger quantity of fluid and much larger percentages of bicarbonates and sulphates were found in the stomach than when a cold solution was administered; furthermore, that the appearance of the acidity and digestive powers of the gastric juice appear much later when a warm solution is injected than after a cold solution. Only, exceptionally, in case VIII, 41, 43, no difference was noticeable in the action of a cold and warm solution of Sprudel Salt. A warm solution, therefore, remains for a longer time in the stomach, and influences its function also for a longer time than a cold solution.

6. The different influence of the soda water as solvent for the Sprudel Salt as compared with distilled water may be observed from the experiments I, 5, 6; II, 13, 14; VI, 32, 33; VIII, 41, 42; XII, 70, 71. From these it will be noticed that when soda water was used in case of alkaline gastric se-

cretion the alkalinity was less; in acid secretion, however, the acidity rose higher and quicker than when distilled water alone was used, so that the maximum of acidity and digestive power is arrived at sooner, but also falls more rapidly in the first than in the last case. The disappearance of the salt from the stomach, as demonstrated by the reaction of the sulphates, is much expedited by the soda water, probably on account of the stimulation to absorption of the free carbonic acid and its power of increasing peristalsis. Soda water increases and at the same time shortens the stimulating action of the Sprudel Salt on the entire gastric function.

SECTION 10.

THE EFFECT OF ADDITIONAL DOSES OF SPRUDEL SALT UPON THE FUNCTIONS OF THE STOMACH.

1. To obtain information on this point, a 5 grm. dose was administered in case X, 54, and aspirated after 2 quarter-hours, the gastric liquid obtained digested and contained no sulphates. If, immediately thereafter, another similar dose was introduced into the stomach (experiment 55), and aspirated after 2 quarter-hours, the gastric liquid obtained entirely failed to digest and contained a large quantity of sulphates. If this was again repeated (X, 53, 56), it was found that, after aspirating 3 quarter-hours later, the liquid obtained, although containing no sulphates, the digestive powers of the gastric juice had not yet been regenerated. It therefore follows that the injury to the gastric function is increased by repeated doses, and the stimulating effect appears reduced. In addition, it may be mentioned that larger doses lower the gastric function much more and are much less stimulating upon the gastric function than small doses. For in case XII, 65 and 66, the gastric liquid, after 5 and 10 grm. doses, was kept at the same reaction, still the gastric liquid obtained after the 5 grm. Sprudel Salt digested quicker and more completely than that obtained after 10 grm.

2. The action of the Sprudel Salt upon the gastric function when used for a longer time differs on the last day from that of the first. Experiments V, 27 and 27 (*b*), as also XIII, 68, 69, of the table, which were made after two weeks' use of the Sprudel Salt under the same conditions, demonstrate that the stomach reacts much more strongly during the first days of the use of the salt than in the following. The acidity in the later days becomes less, the quantity of sulphates in the stomach greater and the digestive power much more reduced.

SECTION II.

CHANGES IN THE ENTIRE GASTRO-INTESTINAL FUNCTIONS AFTER CONTINUED USE OF THE SPRUDEL SALT.

In what way, however, the continued use of Sprudel Salt influences the gastro-intestinal function, I have had the opportunity to observe and closely study, especially in 3 cases. Inasmuch as the individual experimented upon was continually under my control at home and could finally be discharged with decided improvement of 'his diseased condition, I will give the history of his disease in detail.

W. N., æt. 47, married; Catholic; village shoemaker from Krzeszów. According to his own account, he has suffered for the past 4 years from gastric ailments. They manifest themselves generally after eating food that is difficult of digestion, and consist of fullness, oppression, frequently burning and pains in the stomach radiating to the back and chest; in addition there are sour eructations. Concerning vomiting, the patient states that it occurred only 2 or 3 times during his illness, and then consisted of food only. The patient has appetite for all kinds of food, but fears to take any quantity, fearing thereby to cause the gastric pains. The patient describes his pains as being very severe, is much depressed and gives the impression of being a hypochondriac. Had no evacuation of the bowels for several days at a time, the stools then consisting of several balls. The patient also complains of great weakness since the appearance of the disease.

Physical examination revealed the following: The individual is well built, but very moderately nourished; color of skin and face pale. Nothing abnormal could be found in any organ, with the exception that a peculiar succussion in the region of the stomach could be produced during the afternoon, but which was never found early, fasting. The patient was placed upon a simple and uniform diet, lasting during the entire period of treatment and observation. This daily consisted of the following: For breakfast, a half litre boiled milk, two boiled eggs and a roll; for dinner, meat soup, a ration of beef or an allowance of roast beef with gravy and half a roll; evenings at 8 o'clock, a glass of tea with sugar, two soft boiled eggs and half a roll.

The patient was now subjected to an internal examination of the stomach, according to the Krakauer method, the details of which are given in the table under No. XXV:—

(*a*) Examination of the empty stomach (experiment 132) with 100 c.c. distilled water showed a pure gastric juice, containing, however, a large percentage of *HCl* and of intense digestive power.

(*b*) The egg albumen method (experiment 133) showed a similar result in reference to the acidity.

(*c*) The beefsteak method (experiment 136) also shows a high degree of acidity and mechanical weakness of the powers of the stomach.

(*d*) With the gastric volumeter the capacity of the stomach was found to be 3100 c.c., at a water pressure of 22 cm.

(*e*) The microscopical examination of the contents of the fasting stomach showed whole heaps of round, single and double cell nuclei.

This examination led to the diagnosis of an acid disease of the gastric mucous membrane in connection with a dilatation of moderate degree (in comparison with the average stomach capacity of the Polish peasants).

The treatment was begun on the 19th of March, 1884. The Sprudel Salt was administered during the entire period of treatment by means of the stomach tube, in a 5 per cent. watery solution. From the 19th of March to the 5th of April—*i. e.*, during 16 days—10 grm. Sprudel Salt were introduced at 7 A.M. into the empty stomach, and again 10 grm. at 6 P.M. (20 grm. daily). During these 16 days of treatment, the patient had 2 mushy passages the first 4 days; later on, 3 at different times during the day, which afforded him much relief of the abdominal pains. The patient felt, altogether, subjectively better from day to day, had more confidence to eat, and developed a voracious appetite. The former dread of victuals disappeared to such a degree, that the patient occasionally allowed himself to take other food than the above-described diet; but as the deception was brought to light by means of the stomach tube, he submitted to the prescribed régime. From the 6th to the 10th (during 4 days) 30 grm. Sprudel Salt were given daily, 20 grm. at 6 A.M., fasting, and 10 grm. at 6 P.M. Whilst during the previous period the patient felt quite well—so much so, in fact, that he wished to return home—there appeared a change in the subjective symptoms of the patient during these days. He complained of sleeplessness, startling dreams, headaches, feeling of heat in the head (although, objectively, no local increase of temperature could be noticed), general weakness, then followed nausea, eructations after eating, and a disagreeable feeling along the entire œsophagus. Although the passages remained of the same number (2–3 daily), they were more watery. During this second period—*i. e.*, in the third week of the treatment—observations were made of the gastric function early, fasting, before the administration of the salt. They showed the following compared with the condition before the treatment: The acidity and digestive power of the contents of the stomach, fasting, much reduced (table, experiment 137); the same was also shown by the albumen (experiments 138 and 139) and ice-water method (experiment 140); at no time could sulphates be discovered in the fasting stomach from the previous day. Although objectively there was found a beneficial change in the gastric functions, on account of the subjective gastric ailments, the dose of the Sprudel Salt had to be reduced. On the 11th and 12th of April, the dose of 20 grm. Sprudel Salt was again returned to. The general subjective ailments, however, did not subside, and a want of appetite and disgust for food set in ; in addition, there was tenesmus, very watery stools during the day, and severe burning in the rectum, so that to alleviate the intestinal symptoms recourse had to be had to opium suppositories, and the dose of Sprudel Salt reduced to 10 grm. daily, taken at one dose, fasting, from the 13th to the 17th of April. Not until then did the subjective symptoms begin to subside, the number of passages were reduced to 3, and the condition improved to such a degree that, on the 18th and 19th of April, 20 grm. were again given daily (*i. e.*, 10 grm. early, fasting, and 10 grm. in the evening), whereupon the treatment with Sprudel Salt was discontinued. There was a pause made in the treatment with Sprudel Salt from the 20th to the 23d of April, but the same diet, however, was continued. During this time the patient complained of no ailment whatsoever. The following, however, was now noticed, and found to continue until the end of the following period of observation: The original voracious appetite has given place to a feeling of satiety, and the patient, who, at the beginning of the treatment, continually complained of hunger and wanted more food, felt completely satiated with the prescribed diet, so that he occasionally refused the roll at dinner, because he felt no desire for it. Furthermore, the passages did not return to normal, in spite of the fact that the Sprudel Salt was discontinued; there would always be 2-3 mushy passages, accompanied with straining, and this condition continued still for a number of days, during which the patient remained under observation. After an intermission of 3 days, the examination of the gastric function, compared with that before the treatment, showed the following result: The secretion of the gastric mucous membrane has become much less, inasmuch as the gastric liquid aspirated in the first and last experiments had become considerably less.

The acidity of the contents of the fasting stomach is, according to experiment 142, reduced from 13 to 4, the rapidity of digestion from 4 to 15 hours. According to the ice-water method, however (experiment 143), the acidity was reduced from 17 to 3, the time of digestion from 3 to 22 hours. According to the albumen method (experiment 144), however, the acidity was reduced from 42 to 6.5, and the digestion from 4 to 24 hours. The beefsteak method (experiment 141) also showed less acidity (24), which, however, for the most part was dependent upon the fatty acids (drops of grease were seen floating on the gastric fluid). The capacity of the stomach, measured at this time, appeared somewhat smaller (2900 c.c.), in spite of greater height of the water column (26 c.c.), *i. e.*, the vital contractility of the gastric walls has become somewhat stronger owing to the treatment. In reference to the mechanical power, it must be admitted that it also had become improved, for, in spite of the reduced chemism of digestion, less undigested albumen, as also fewer particles of meat, were found by the albumen,* as also by the

---

* It occasionally happens that with the albumen method, even in a chemically normal stomach, isolated pieces of albumen will be brought to light, even after two hours, if the stomach is carefully rinsed; these, however, have sharp and entire margins, and have probably been retained in the folds of the mucous membrane; if they are in addition tinged

beefsteak method, than were found at the beginning of the treatment. The patient was now allowed to return to his home without any subjective ailments whatsoever, and with the conviction that he was cured.

The second case examined in the same direction was S. B., æt. 23, village servant. He came to the clinic on June 3d, on account of violent subjective gastric ailments. External physical examination showed nothing abnormal, the internal examination of the stomach, however, executed in precisely the same manner as in the above case, showed the following: On June 4th the clear gastric contents of the fasting stomach of the degree of acidity 22, with very intense hydrochloric acid reaction, and containing 32 per cent. of Cl. The artificial digestion finished after $3\frac{1}{2}$ hours, peptonization complete. On June 5th the albumen method also yielded a clear gastric fluid of the acidity 22, extremely intense HCl reaction and 38 per cent. Cl. The artificial digestion completed after $3\frac{1}{2}$ hours, peptonization complete. On June 6th the capacity of the stomach was 2500 c.c., with the water column at 12 cm. The patient was ordered to take a tablespoonful of Sprudel Salt in two glasses of warm water, fasting. He only took the same for eight days, then remained a whole week without medicine, until he again came up for examination on June 22d. He said that he felt better and could now eat more, because the constriction and burning in the region of the stomach was much less. The internal examination of the stomach, made on June 22d, showed the following: The clear contents of the stomach obtained, fasting, showed a degree of acidity of 11, an intense HCl reaction and 31 per cent. per c.c. of Cl; the artificial digestion was finished in five hours. The ice-water method, performed on June 23d, showed the acidity to be 18 degrees, very intense HCl reaction, 30 per cent. per c.c. of Cl. Digestion was completed in four hours. On June 24th the capacity of the stomach was 2300 c.c. at 11 cm. water column. It will, therefore, be seen that here also, after but a short use of the Sprudel Salt, a tendency to a reduction of the acid secretion and lessening of the gastric ailments was produced.

A similar state was also found in the individual cited under XIX of the table. The ordinary physical examination revealed no cause for the gastric ailments of the patient. The internal examination of the stomach, made before the beginning of the treatment with Sprudel Salt, showed the following: The contents of the stomach, fasting, was clear, acidity 12 degrees, hydrochloric acid reaction strong, artificial digestion completed in 4 hours, peptonization complete. The albumen method showed even after 6 quarter-hours the stomach contents to contain traces of flocculent pieces of albumen, acidity 22, intense HCl and peptone reaction. In addition there was obstinate

---

yellow, then it may be assumed that they have returned from the duodenum into the stomach. The termination of digestion is not demonstrated so much by the absolute disappearance of the pieces of albumen as by the rapid reduction of the acidity of the gastric fluid to normal.

To obtain uniform results from the internal examination of the stomach, it is advantageous to introduce the necessary diluent (100 to 200 c.c. of water) five minutes before aspiration.

constipation lasting several days, in consequence of which the patient complained of oppression and weight in the abdomen. The patient was ordered two teaspoonfuls of Sprudel Salt, each dissolved in a glassful of warm water, to be taken at intervals of one hour. He continued their use during two weeks, taking them early, fasting, and had during this time two to four mushy passages during the day. After discontinuing the salt there was one passage daily; the patient declared that his abdomen was relieved; had no oppression and gastric flatulence, but felt a sense of warmth in the region of the stomach, belched from time to time and felt weak. After having discontinued the salt for ten days the patient was subjected to an internal examination of the stomach, which demonstrated the following: The contents of the fasting stomach clear as water, 7 degrees acidity, distinct HCl reaction, artificial digestion completed in five hours, peptonization pretty distinct. The albumen method showed the contents of the stomach, after 6 quarter-hours, to be clear as water, barely containing traces of flocculi, but of an acidity of only 2 degrees, no HCl reaction, no pepton nor albumen reaction; artificial digestion only completed after $4\frac{1}{2}$ hours, with the help of HCl. The gastric digestion was, therefore, almost completed in the proper time.

In this case the acid secretion was already much reduced after two weeks' use of the Sprudel Salt, the gastric digestion and the evacuations of the bowels brought to nearly normal, and most of the subjective gastric ailments removed.

If these three instructive cases are more closely analyzed, it will be seen, especially from the first carefully observed case, that the long continued (one month) use of larger quantities of Sprudel Salt had entirely altered the gastro-intestinal functions. The excessive acid secretion was changed to acid insufficiency, the too energetic digestive power was much reduced, without at the same time weakening, but rather increasing the mechanical power. The stomach capacity and the gastric contractility have also been benefited. The condition of the stomach which was found by the treatment with Sprudel Salt, recalled the characteristics of mucous gastric catarrh, inasmuch as the last aspirations of the gastric contents contained much gelatinous mucus, which was not observed in any case before the treatment; and I am of opinion, judging from examinations with Carlsbad water, that, if in this case the experiments had not been stopped, but the treatment continued for two more weeks with Sprudel Salt or some other saline remedy, a mucous catarrh with total loss of the digestive chemism would have been produced. Although one can never judge of the functional or anatomical condition of the stomach by the subjective gastric ailments, still attention must here be called to the change in many feelings before complained of. The ravenous appetite, which was only held in check by the fear of the pain following the ingestion of food and the empty feeling about the stomach, gave place to a feeling of satiety and fullness after the acidity had been reduced. Furthermore, in reference to the functions of the intestines, the habitual obstinate constipation was changed into diarrhœa (in the third above described case to almost normal). This is the more noteworthy, because it is generally assumed in practice that the salines, when given in large doses for a longer time, are followed by obstinate constipation.

The very important fact, which I have verified in reference to the Carlsbad water by experiment, is, therefore, also true of the Sprudel Salt, that, when continued for a time *it lowers the digestive chemism, the increase or exacerbation of which, according to my observations, is the cause of the majority of gastric disorders. It accomplishes this, however, without enfeebling the mechanism.* If, however, the change in the function of the intestine, which was noticed in this case, was to be sought in the exacerbation of the intestinal mechanism, or in the stimulation or irritation of the intestinal mucous membrane, cannot here be more definitely stated, on account of the want of methods for examining the intestinal functions. I feel disposed to assume the latter to be the case, on account of the fact that large quantities of Sprudel Salt exert a local action even on the lowest portions of the intestinal tract, there causing burning and even capillary hemorrhage. I have used in my experiments in the first case above cited, the largest quantities of Sprudel Salt, and in the most concentrated solution that it is possible to use in man. This was necessary and possible, because this case belonged to the category suffering from the greatest degree of acid hypersecretion and constipation at the same time. In less advanced cases, of course, smaller doses would suffice. I would also call attention to the fact that, by the method here mentioned by me, the influence of single as also repeated doses of every medicine upon the function of the stomach may be studied directly upon the human being, and will, therefore, lead to more useful results than if they were determined by experiments upon animals.

I am much less able to report anything definite concerning the behavior of the Sprudel Salt when long continued in cases of gastric acid insufficiency. In this direction I have had the opportunity to observe but few cases in the above manner. I have generally, in such cases, given 5 grm. Sprudel Salt in $\frac{1}{4}$ litre soda water cooled by means of ice, and directed it taken early, fasting, for a number of days. Of three such cases there was in one case a distinct alleviation of the subjective symptoms; in two other cases there was no improvement. Objectively there was found in the case that improved, disappearance of the organic acids, appearance of slight traces of hydrochloric acid and artificial power of digestion, as also the more rapid disappearance of the albumen from the stomach. The facts of the case must, however, be more fully established by additional cases, so as to ascertain if small doses of Sprudel Salt are able, the same as Carlsbad water, to improve the chemism and mechanism of digestion in acid insufficiency.

SECTION 12.

THE TOTAL RESULT OF THE INFLUENCE OF SPRUDEL SALT ON THE ENTIRE GASTRO-INTESTINAL FUNCTION.

If the individual, partial results obtained be summed up as a whole, so as to delineate a collective idea of the action of the Sprudel Salt on the whole intestinal function, the following will be seen:—

1. If moderate doses (5–10 grm.) Sprudel Salt, in watery solution, be introduced into a stomach deviating but slightly from the normal, the gastric acid and digestive ferment are primarily totally destroyed; the digestive property of the gastric contents are completely arrested. At the same time the mucus is dissolved and other morphotic constituents and fermentation organisms are converted into a semi-soaked condition and appear suspended in the contents of the stomach. Already in the first quarter of an hour the salt introduced proves itself a powerful stimulant to call forth the mechanical and chemical functions of the entire digestive canal. The excited mechanical activity of the stomach manifests itself by passing the saline solution on into the intestine, by the presence of bile in the stomach, by occasional nausea and eructations, and in nervous individuals even by vomiting of the salt. The movements of the stomach soon start the intestinal peristalsis, for already in the first quarter-hour rumbling is produced in the abdomen and intestinal gases voided. The intestinal peristalsis soon increases, manifesting itself by a desire to go to stool, and in a short time, 1–2 quarter hours, the lowermost section of intestine is emptied of fecal masses. The lower the saline solution penetrates from the stomach into the intestine, the stronger the stimulation to intestinal movements becomes, more liquid and gaseous contents are produced, mixed with saline solution in the large intestine, and produce in consequence liquid passages strongly colored with bile and having a penetrating odor, and these may be followed when very large quantities of salt are used by additional watery passages coming from the uppermost portions of the intestinal canal, which contain only salt, and cause the burning. Such an excitement of the intestinal tract, lasting a number of hours, is followed by a relaxation of the mechanical powers, and although the salt still fills recesses of the intestine, which manifests itself by the succussion sound, it remains quiet and the evacuations cease. It may therefore be assumed that the alvine evacuations following the Sprudel Salt are rather dependent upon increased peristaltic action than upon profuse mucus secretion, and that the body of the same consists of the contents of the entire intestinal tract mixed with unabsorbed saline solution. The following circumstances prove this: (*a*) The rapid manifestation of the action of the salt upon the movement of the whole intestinal tract already in the first quarter-hour, whilst it is still almost wholly present in the stomach. (*b*) The rapid passage of the gastric contents into the stools (the saline solution

may appear in the evacuations in half an hour). (*c*) Occasionally, in spite of the large quantity of salt taken, only one or two solid passages are produced, owing to the excited peristaltic action, the saline solution, however, remaining for a long time in the intestinal tract on account of the relaxation of the intestinal mechanism, which always comes on near the end of the evacuations, allowing the solution to remain until it is slowly entirely absorbed.

The stimulation of the gastro-intestinal tract to chemical action by the salt is synchronous with the above-described mechanical action, but is restricted in the main to its upper division, and consists in the stimulation of the secretion of gastric juice and of the secretion of bile. An abundant acid secretion is produced in the stomach, which, however, is made latent by the action of the alkalies (the first stage), until these latter are neutralized by the gastric acid. This takes place, as a rule, in the third quarter-hour after ingestion of 5 grm., and at the end of the second hour after the ingestion of 10 grm. Sprudel Salt. After this period the secretion of acid progresses still further (stage of excitation), probably owing to the stimulation of the small quantity of salt still present in the stomach. After a certain time the secretion of acid reaches its maximum, which is higher and reached more rapidly after 5 grm. than after 10 grm. From here on the acidity decreases more rapidly (stage of decrease of reduction) than it had previously increased, and that to below that of the empty stomach, so that during a certain time there will be a period of acid insufficiency (lassitude).

All these stages run their course within 2–3 hours. The digestive ferment, which had been destroyed by the Sprudel Salt, is regenerated in a similar way and synchronous with the gastric acid, with the difference that the regeneration of the same proceeds much slower than the gastric acid, so that the time of greatest efficiency of the gastric juice is later than the maximum of acidity. The original digestive power, however, which existed before the experiments does not return for several hours.

As soon as the saline solution enters the duodenum, which certainly occurs in the first quarter-hour, a stimulation of the same and of the biliary passages is produced, from which, most probably, the accumulated bile is poured copiously, and finds its way, for the most part, into the stools, but not infrequently also into the stomach.

The absorption of the constituents of the saline solution progresses parallel with the stimulation of the mechanism and chemism of the gastro-intestinal functions. The sulphates have disappeared from the stomach in 1–1½ hours, and are swept away, for the most part, with the fluid passages. The carbonates, however, are not to be found in these; they may, however, be detected in the urine in 1 hour, and render it alkaline, so that the sodium bicarbonate may be looked upon as the constituent of the Sprudel Salt that is absorbed in greatest quantity.

Two small doses of Sprudel Salt, up to 5 grm., stimulate only the function of the upper portions of the digestive canal; larger doses, over 10 grm., are exhausting and harassing upon the function of the upper portion and strongly stimulating upon the lower portion of the intestinal canal. If, however, they are given in divided doses, the exhaustion of the function of the upper portion is much less. The lassitude only follows the subsequent repeated doses, whilst the first are stimulating.

When long continued, the stimulating action of the Sprudel Salt upon the gastro-intestinal function becomes less and less, until it finally ceases entirely.

3. The influence of the Sprudel Salt upon the functions of the stomach, when long continued, consists of the reduction of the acid secretion and digestive power, with, at the same time, no alteration or increase in mechanical power, which shows itself in decreased sensitiveness of the organ against chemical and mechanical irritations. A continued use of large doses of Sprudel Salt is accompanied with lowering of the general nutrition and subjective changes in the general condition (feeling of weakness, dullness of the head). The anatomical causes of these functional changes in the stomach cannot be determined by our present knowledge of the connection between the functional and anatomical lesions of the stomach. In what sense the intestinal function is changed by the continued use of the Sprudel Salt I can only refer to my conjectures in § 11.

Small quantities of Sprudel Salt continued for a certain time in some not well-defined cases of gastric acid insufficiency act as a stimulant to the acid secretion and the gastric mechanism.

Sprudel Salt may, therefore, be called a "stomach salt" *par excellence*, which, in a single small dose, acts as a stimulant; when long continued, however, lowering the chemical functions of the stomach, while the former old-fashioned *crystallized* Sprudel Salt principally acts upon the intestine, and may, therefore, be designated a purgative.

4. From the foregone the following clinical observations and facts are easily understood :—

(*a*) That all remedies which stimulate the function of the stomach and bowels, given together with the Sprudel Salt (powder form) increase its action. Therefore, soda and seltzer water used as solvents of the Sprudel Salt are more powerfully stimulating upon the gastric chemism and intestinal mechanism, than an ordinary watery solution of the same. In the same way does exercise act, after the ingestion of the salt, in promoting evacuation of the bowels.

(*b*) That the action of the Sprudel Salt takes place only at certain intervals momentarily and lasts as long as the intestinal peristalsis continues, and that, in case the patient misses this intestinal movement, which manifests itself in a desire to go to stool, the passage fails to follow, although the bowels are completely filled.

(*c*) That in nervous individuals the Sprudel Salt exerts a greater influence upon the function of the bowels than in torpid individuals.

(*d*) That in irritated conditions of the large intestine the Sprudel Salt which reaches there calls forth a feeling of burning and of pain.

(*e*) That in cases of constriction of the pylorus, as also in case of partial impermeability of the intestinal canal after introduction of the Sprudel Salt, abdominal distress, pain and griping supervene, which may be ascribed to the collection of the saline solution and the irritation of the same in one place.

III.—CLINICAL DEDUCTIONS BASED UPON THE EXPERIMENTS MADE.

I now come to treat of the therapeutic use of the Sprudel Salt, based partly upon the material presented and partly upon experience in private practice. Inasmuch as the individuals experimented upon, as well as the experience of physicians heretofore in the use of the new Carlsbad salt do not quite suffice to establish positive indications for its use, I will first treat of the much more positive theme of the restriction in the use of the Sprudel Salt.

SECTION 13.

THE RESTRICTIONS AND CONTRAINDICATIONS IN THE USE OF THE CARLSBAD SPRUDEL SALT.

As the Sprudel Salt is not a remedy which endangers life, it is self-evident that there are no absolute contraindications for the same. Under the head of contraindications only such cases can be meant in which the pathological condition is not ameliorated or even made worse, or where, on account of the symptoms caused, the use of the Sprudel Salt is not deemed advantageous.

1. Above everything it must be emphasized that the systematic use of large doses, above 15 grm. (3 teaspoonfuls), given daily for some time, is not advisable; inasmuch as such large doses, as has been demonstrated by the experiments, completely lower the digestive chemism of the stomach, arrest the resorptive power of the small intestines, cause a mucous gastric catarrh, and occasion a condition of irritation of the large intestine. General nutrition thereby suffers very much, and the patients feel subjectively very bad. If it were desired to continue the use of a larger dose than 15 grm. the rest would have to be given per rectum, which, however, could not be long continued on account of the irritation of the intestine from above and below.

2. In cases of insufficient gastric secretion, larger doses than 5 grm. Sprudel Salt given continuously are always to be avoided, as also the taking of the solution warm. There is danger in these cases of complete suppression of the acid secretion and of hastening the production of the mucus secretion.

3. In cases of great dilatation of the stomach, in which the much delayed emptying of the stomach manifests itself by the retention of remnants of food until the following day, or in cases of stenosis of the pylorus, it is not justifiable to attempt to force a movement of the bowels by larger doses of Sprudel Salt. The greatest subjective abdominal ailments are caused thereby, because the solution of salt can neither be quickly emptied into the intestine nor be absorbed, irritates the walls of the stomach and destroys the secretory elements of the gastric mucous membrane, and still, in the majority of cases there is no evacuation of the bowels produced.

4. In all diseased conditions of the large intestine which are still accompanied by very intense irritation, as, immediately following acute enteritis, dysentery, acute proctitis, as also in ulcerated conditions of the rectum, large doses of Sprudel Salt dare not be used, as there is danger of even detaching pieces of diseased mucous membrane, occasioning hemorrhages and subjectively very painful tenesmus and burning during defecation.

5. In partial impermeability of the intestinal tract, the bowels dare not be forcibly moved by the systematic use of Sprudel Salt, inasmuch as the solution of the salt accumulates in the section of intestine in front of the stenosis and frequently causes very disagreeable subjective gastric ailments.

6. In cases of very severe constipation, the systematic use of Sprudel Salt as a purgative, given per os, is not to be recommended, because the large doses necessary in these cases are followed by the ailments enumerated under 1.

7. In persons who are much emaciated, or who suffer from great nervousness or have a valvular disease of the heart without compensative hypertrophy, it is best to abstain from the systematic use, especially of large quantities, of Sprudel Salt by the mouth, because the lowering of the general nutrition and weakness, and thereby the nervousness, become much greater.

8. Although the use of Sprudel Salt by the rectum is permissible in all diseases excepting those enumerated under 4, still it is necessary to guard against the systematic use of too concentrated solutions of the salt. The maximum of concentration dare not exceed 5 per cent.

SECTION 14.

INDICATIONS FOR THE USE OF SPRUDEL SALT.

In reference to the clinical indications for the use of Sprudel Salt, I can only give positive data of such as concern the pathological conditions of the gastro-intestinal canal; for the others, however, I can give more or less well founded conjectures. I must here observe that the indications of the Carlsbad Sprudel Salt (powder form), which is a combination of a number of different salts, are entirely different from those of the former Sprudel Salt (crystallized), which is composed almost entirely of sulphate of sodium. Inasmuch as the Sprudel Salt (powder form) is only of recent introduction, and accurate practical experience is still wanting concerning it, it is difficult to set up infallible indications based on experience. The chief pathological conditions in which Sprudel Salt may be used with advantage and success are, in the main, those which call for the use of the Carlsbad Thermal Waters, whereby, however, that must always be borne in mind which is given in the previous chapter under the head of contraindications.

1. Above all, and looked upon by myself as a specific at the present time, is the Sprudel Salt in the treatment of *simple acid and catarrhal gastric hypersecretion*, a disease which lies at the foundation of at least one-half of all the gastric ailments, and especially amongst the male portion of the Polish Israelites. In these cases medium doses (10 grm.), and in greater hypersecretion larger doses (15 grm.) must be given during a long period, systematically. The solution of Sprudel Salt must be taken warm (40° C.) and somewhat concentrated (3 per cent. to 5 per cent.). It is desirable to have the solution remain as long as possible in the stomach, and for the salt to act energetically upon the mucous membrane, this being best accomplished by the above procedure. As nervous individuals, as, for instance, most Israelites are, cannot tolerate large doses of Sprudel Salt for a long period, it is better to begin the treatment with moderate (10 grm.) doses.

2. *In insufficiency of the gastric acid*, which is much less frequently met with in practice, Sprudel Salt may also be used with a view to its stimulating effect. In these cases, however, only small doses, 5 grm. at the most, may be given in a perfectly cool solution, preferably made with carbonic acid water, and the patient must be warned not to take larger doses on his own responsibility. And should it become necessary to produce evacuation of the bowels, it is better to accomplish this by using the salt in the form of an injection by the rectum, than to give it by the mouth. All cases of insufficiency of acid secretion cannot, however, be cured in this way; if the gastric acid does not become regenerated, a further continued use of Sprudel Salt is unavailing and even harmful. Therefore, when the internal gastric examination, after the continuous use of the Sprudel Salt for two weeks, does not show an increase in the acidity, but rather a decrease of the same, it

is best to discontinue the use of the Sprudel Salt. In these cases it may probably be assumed that the secretory apparatus of the gastric mucous membrane has undergone a retrogressive anatomical metamorphosis.

3. The *passing dyspepsia*, caused by excesses in the use of alcoholic beverages, and generally known as "swelled-head," subsides after the ingestion of a small (5 grm.) dose, dissolved in right cold or carbonic acid water, and taken early, fasting, and is followed by a most excellent appetite. A dyspepsia caused by overloading the stomach requires larger doses (10–15 grm.), also given in a perfectly cold solution. In these cases the action of the salt on the bowel must also be made to manifest itself.

4. In *gastric ulcer*, the old-fashioned crystallized Sprudel Salt has been recommended in various quarters, especially by Leube, given in a warm solution (1 teaspoonful to ½ litre of lukewarm water). I think that all that has been said in favor of the crystallized is still more true of the Sprudel Salt, powder form. I would only suggest that the Sprudel Salt, powder form, be given several days after the arrest of the hemorrhage, for fear of loosening the thrombus by the alkaline solution and thereby renewing the bleeding.

5. For the purpose of *washing out the stomach in cases of gastric dilatation*, accompanied with weakened muscular power or with stenosis of the pylorus. The purpose in these cases is to remove the mucus, fermentation organisms and the putrefying food remnants; to arrest fermentation, as also to reduce the acid hypersecretion which is so often present. The larger quantities of Sprudel Salt which would be necessary for this purpose could not be given by the mouth in these cases without aggravating the gastric ailments. Nothing remains but to remove the gastric contents and to wash out the cavity with a solution that at the same time acts therapeutically upon the mucous membrane. For this purpose a solution of Sprudel Salt is well adapted, inasmuch as it dissolves the mucus, causes the morphotic constituents and ferment organisms to swell up and acts well upon the mucous membrane.

These five indications stamp the Carlsbad Sprudel Salt (powder form) as a true stomach salt.

6. As an *occasional purgative* the Sprudel Salt belongs to the best-disposed of these remedies, and may be used in all cases where the contraindications mentioned in the preceding chapter do not exist. As to the size of the dose, the physician must be guided by the history of the case, by inquiring as to the dose of other purgatives or mineral waters which the patient is in the habit of taking, or if the patient be inclined to diarrhœa. Although the medium dose is 10 grm., it will be found too large in nervous individuals and too small in very torpid ones. As a purgative the Sprudel Salt is best given dissolved in spring or soda water.

7. In *habitual constipation* of moderate degree the Sprudel Salt (powder form) will be found to be more successful even than the Carlsbad thermal waters themselves. The quantity of Sprudel Salt in these cases, when

systematically used, should not exceed 10 grm.; in fact, had better be less. The solution must be drank cold. In case larger doses are necessary it is advisable to order them taken in soda water, or better still, in a carbonated mineral water, as Szczawnica, Giesshübler or Krondörfer water. By the use of these solvents the dose of the Sprudel Salt may frequently be reduced to 5 grm., because of the somewhat stimulant action of the mineral water itself upon the intestine. In very severe cases of constipation a part of the salt may be given by the rectum.

8. In *intestinal catarrh* accompanied with constipation. In these cases the object is to remove the masses of mucus, to stimulate the mucous membrane and muscular coat of the intestines. These indications are well supplied by the Sprudel Salt, in fact, even better than by the Carlsbad water, as the latter does not reach the lower portion of the bowels, and hence cannot produce any local action there as the Sprudel Salt does, which even passes out with the stools. In intestinal catarrh, medium (10 grm.) doses at least should be given, and those in a lukewarm solution, as this remains longer in the intestinal canal, and can therefore act better locally. But as large doses when long continued are generally not well borne, and lower the function of the stomach, it is always more advantageous, and also more adapted to the nature of the pathological condition, to allow at least part of the Sprudel Salt to act locally by means of an injection. At least in the first days of the treatment copious, warm injections of Sprudel Salt are to be ordered; they should wash out the intestine as high up as the cæcum. Later on, smaller injections may be ordered, which the patient is to retain during the whole night. The concentration of the injections for the purpose of washing out the mucus is at first 3 per cent.; later on, however, 1 per cent.

9. *Catarrhal Jaundice.*—In most of these cases the desideratum is to remove the obstruction in the bile ducts and biliary passages. Inasmuch as, according to the experiments cited, the Sprudel Salt very much favors the evacuation of the bile, obstructive jaundice is a very good subject for treatment with Sprudel Salt. The dose used must be larger, 10–15 grm., and, so as to get the solution to remain in the upper portion of the intestinal canal as long as possible, the solution must be drunk warm (30–40° C.), and the treatment continued systematically for at least 14 days.

10. *In chololithiasis, or gall-stones*, success may also be expected from the use of the Sprudel Salt, which may be termed a concentrated Carlsbad water; a case successfully treated is described in Section 6, case XXXIX. The mode of administration must be similar to that in catarrhal jaundice, but as it probably, in these cases, would be an advantage to give even larger doses of the salt, a part of the same may be given by the rectum at bed-time. The injection, per rectum, of simple water in diseases of the liver, as practiced in Mossler's clinic at Greifswald (E. Peiper, *Zeitschr. f. klin. Med.*,

Bd. IV, p. 403), were accompanied with very favorable results; so much the more may be expected of the use of a solution of Sprudel Salt.

11. *In fatty liver and in the first stage of cirrhosis of the liver*, the desideratum is to accelerate the circulation of the portal system and the flow of bile; this is accomplished effectually by the use of Carlsbad water, hence success may also be expected from the use of the Sprudel Salt. It also appears advantageous, in addition to the administration of moderate doses (up to 10 grm.) of the Sprudel Salt by the mouth, to at the same time administer some by the rectum, inasmuch as the solution can by this means more readily reach the portal system.

12. *In general adiposis*, or excessive accumulation of fat throughout the system, the desideratum is to reduce the general nutrition, or rather to use up the deposited fat and prevent its proliferation. The Sprudel Salt, given in large doses, accomplishes this result, and at the same time prevents a new deposition of fatty tissue by materially reducing the peptonizing and resorptive power of the gastro-intestinal canal. Therefore, in the treatment of adiposis, the Sprudel Salt may be used, accompanied with a proper diet. It must be given in large doses, 15-20 grm., daily, dissolved in lukewarm water. It would be advantageous to have the patient take his meal soon after taking the Sprudel Salt. By so doing he will be satisfied with less food, and the nourishing properties of the food would be reduced to a minimum by coming in contact with the salt in the small intestines, where it would reduce the peptonization and absorption to a minimum.

Until recently, the crystallized Sprudel Salt, which may be looked upon simply as a brisk purge, was used by the profession, but now, when a salt is being obtained from the Carlsbad water, which is a blending of so many salts, and which may be looked upon as, in addition to being a purgative, to having other effects peculiar to itself, and really representing the mineral water from which it is obtained, clinicians may now use the same in all those cases in which the Carlsbad water has proven itself a specific, or in which alkaline muriatic mineral waters are indicated as:—

13. *In Diabetes Mellitus*, where the Carlsbad water arrests the formation of sugar in so many cases. On account of a want of material the mode of use of the Sprudel Salt in these cases is not given in detail. It may, however, be stated that warm solutions are to be preferred.

14. *In Pyelitis, Cystitis, and Renal Calculi.*—In these cases a warm, not too concentrated solution, which acts as a diuretic and renders the urine strongly alkaline, would be of signal service.

15. In some cases of bronchitis, to favor the secretion of mucus, the Sprudel Salt should be of great service, because of its large quantity of sodium bicarbonate and sodium chloride, at the same time reducing the venous engorgement by its action on the bowels.

It is, therefore, to be hoped that the Sprudel Salt will, in the course of time, be subjected to further clinical research, and that its indications and use will be still further extended.

MODE OF USE OF THE SPRUDEL SALT (POWDER FORM).

The Sprudel Salt (powder form) is used by the mouth, by the rectum and to wash out the gastro-intestinal canal.

SECTION 15.

APPLICATION OF THE SPRUDEL SALT BY THE MOUTH.

1. The most judicious time to administer the Sprudel Salt is the morning, on an empty stomach, in fact, as soon as the patient awakes. In gastrointestinal diseases this is the only time admissible, for in these hours the salt can act upon the quiescent organ entire and, without alteration by other influences, unfold its action. Also in diseases of distant organs the early morning is the most appropriate time, because at this time the several constituents of the salt do not experience any abnormal changes in the upper digestive passages, are best absorbed and passed into the circulation. They then produce their action upon an organism that is affected the least by external influences (as is only the case in the morning). If the Sprudel Salt is introduced into a full stomach during the course of digestion, the entire digestive activity will be at once arrested, and when medium doses are used the food is passed into the lower portions of the intestine, there, by irritation, causing subjective abdominal ailments; when larger doses of the salt are used it is passed out with the stools. If it is desired to evacuate the bowels when the stomach is full, it is, as a rule, necessary to give a dose twice as large as would be necessary in an empty stomach. If, exceptionally, it is desirable to use the Sprudel Salt during the day, it is judicious to wait until 6 to 7 hours after eating.

2. *Dose of the Sprudel Salt.*—A single small dose of Sprudel Salt is a teaspoonful (5 grm.), a medium dose two teaspoonfuls (10 grm.), and a large dose 3 teaspoonfuls (15 grm.). As a rule, only one teaspoonful of the salt dissolved in a goblet of water is taken, this is equal to a concentration of about 2 per cent. Solutions of greater concentration than 5 per cent. are not to be used. Medium and large doses are divided into 2 or 3 small ones, as a single large dose would require either a larger quantity of liquid, or else would be introduced into the stomach in a too concentrated solution. In both cases, as is seen by the experiment, the gastric function is too much affected; in addition, more concentrated or larger quantities of Sprudel Salt solution introduced into the stomach cause nausea and vomiting.

3. *The Solvents.*—Sprudel Salt is always given in solution, generally in

distilled rain-, river- or spring-water. The use of spring water is, however, not advantageous, as a part of the sodium carbonate is converted into calcium carbonate and makes the whole liquid cloudy. A solvent very much to be recommended, inasmuch as it assists the action of the Sprudel Salt and improves the taste, is soda water, also Kissingen, Szczawnica, Bilner and Krondörfer waters. These solvents are, however, intended for making cold solutions, and may be prepared by placing the dose (5 grm.) of salt to be taken into a tumbler, add thereto ¼ glassful of mineral water and dissolve the salt, immediately before taking the dose the tumbler is filled with the mineral water. A *cold solution of Sprudel Salt* is made most judiciously by placing the requisite quantity of Sprudel Salt in a glassful of the solvent in the evening and allowing it to stand over night outside of the window. As soon as the patient awakens in the morning he has a well-cooled solution of the salt, the undissolved portion of which immediately disappears upon agitation. A *lukewarm solution* may have a temperature of 36° C., and a *warm* solution of 50 to 55° C. Solutions having a higher temperature must not be used, as they affect the gastric mucous membrane too much and cause changes in its structure. A warm solution generally acts less upon the bowels than a cold one. A warm solution of Sprudel Salt must not be made by first placing the salt in cold water and then warming the same; the salt must be dissolved in water already warmed, as the sodium bicarbonate would be partly decomposed by the heating and converted into the caustic monocarbonate.

4. *Intervals between Doses.*—A certain time must be allowed to elapse between the doses when the dose of Sprudel Salt is to be repeated. If the individual dose be 1 teaspoonful of the salt the intervals between the doses will be governed by the following: The immediate effect of the alkalies of a 5 grm. dose lasts 3 quarter-hours, for in the third quarter-hour they are neutralized by the gastric acid. Every following dose acts longer on the gastric mucous membrane and affects the gastric function more than the preceding; doses given at shorter intervals produce evacuation of the bowels quicker than when given at longer intervals. Inasmuch as in gastric diseases it is desirable that the action of the salt be as long continued as possible, longer intervals are indicated in these cases, the most judicious being three-quarters of an hour. If, however, the acid hypersecretion in the stomach be great, a 5 grm. dose repeated in half an hour will answer the purpose very well. An interval of half an hour between two doses of 5 grm. is also the general rule for all other cases. If 15 grm. be given in three divided doses, the first interval may well be half an hour, but the second must be at least three-quarters of an hour, so as not to affect the gastric function too much or to paralyze the intestinal functions. If a rapid evacuation of the bowels be desired the first interval is best shortened to one-quarter of an hour.

5. *The Behavior whilst taking the Sprudel Salt.*—The conduct of the patient whilst taking the Sprudel Salt should be the same as when drinking

a mineral water. As exercise favors the evacuation of the bowels, the patient does well to take some exercise in the open air during the intervals; but he must be enjoined not to suppress the desire to go to stool too long, but attend to it at once, or else the evacuation will not take place at all.

SECTION 16.

THE DIETETIC RÉGIME DURING A TREATMENT WITH SPRUDEL SALT.

(*a*) The first question in this direction is, *When, after taking the Sprudel Salt, shall breakfast be eaten?* Here the consideration must not be lost sight of, that the Sprudel Salt be allowed to act uninterruptedly as long as possible. This cannot, however, take place if the intestinal canal is made to exercise its digestive function by means of food, and the solution of Sprudel Salt is chemically changed by being mixed with the food. It is best to allow the salt to exert its action as long as possible. This is, primarily, especially necessary in diseases of the stomach. As the action of the salt upon the gastric mucous membrane does not cease until the third stage, the stage of depression, during the treatment of gastric diseases by means of Sprudel Salt, no food should be placed into the stomach until this period. Although the stage of reduction or depression sets in at different times, according to the case and dose given, as a rule, however, it may be said to set in in $1-1\frac{1}{2}$ hours, and after a 10-grm. dose after 2 hours. Breakfast must, therefore, be postponed to $1\frac{1}{2}-2$ hours after taking the Sprudel Salt in gastric diseases, and in some cases, as in dilatation, even longer. If the food is taken sooner, then not only is the full action of the salt lost, but also the proper beneficial effect of the nourishment, for the peptonization and the power of absorption of the entire gastro-intestinal tract is much weakened; in addition to this, undigested remnants of food are driven into the lower portions of the intestines by the stimulant effect of the salt upon the peristalsis, where they produce conditions of irritation and cause subjective ailments; this circumstance is the more critical when the gastric ailment is complicated by an intestinal disease. It is, therefore, readily understood why patients who take saline remedies or mineral waters for a long time do not lay on flesh, in spite of the quantity of food taken. In all other cases of disease, also, the physician must endeavor to keep the patients, who frequently have a voracious appetite after taking the Sprudel Salt, from taking breakfast as long as possible. Only in one case would it be indicated to allow the patient to take breakfast at the end of the first hour, so as to satisfy the feeling of hunger and to prevent the patient getting the full benefit of the food as much as possible; this is the case in the treatment of corpulency with saline remedies or mineral waters.

(*b*) The next question concerns the *quality and quantity* of nourishment during a systematic treatment with Sprudel Salt. Even though the diet must always be adapted to the individual and to the disease, still, it is best to have a certain general standard or rule to follow. This may be derived from the following considerations: During the use of the Sprudel Salt the entire intestinal tract is thrown into a condition of irritation, the peptonization and power of absorption is weakened for the first half of the day, and by the continued use the general nutrition is much lowered. On account of these circumstances, as also not to do harm by the diet in the special case, but to assist the therapeutics as much as possible, it is well to use a moderate, unirritating but strong and easily-digested diet, as is done in Carlsbad. It is, therefore, necessary to abstain from all chemically-irritating articles of food and condiments, as fat and very greasy food, strong spices, all strong alcoholic beverages; also mechanically-irritating food, tough meat, raw fruit, all foods that leave a great quantity of indigestible residue in the intestines and remain too long in the gastro-intestinal canal, having a tendency to undergo fermentation and keep up the condition of irritation of the intestinal mucous membrane (most vegetables and compotes, farinaceous articles of food known to be difficult of digestion, large quantities of any starchy food); lastly, all articles of diet, even those considered very easy of digestion, which are known, from the history of the patient, to disagree with him. As regards sour foods and acid drinks, it is best to abstain from the same, as nothing definite is known of their effect on the gastro-intestinal tract; at the most, something acidulous may be allowed in the evening; this holds good especially in gastric acid deficiency. The daily quantity of food must also be regulated, so that the main meal come at noon or during the afternoon—*i. e.*, not until the irritation of the intestinal canal has subsided and its digestive and resorptive power been again restored. Therefore the following may be taken as a bill of fare during treatment with Sprudel Salt: At breakfast, according to habit, tea, milk or coffee, with little bread (at 11 o'clock, in case of excessive hunger, 2 soft-boiled eggs); at 1 o'clock, dinner, consisting of beef soup with grits, or bouillon, roast meat without fat (veal, beef, chicken, deer), beefsteak, or chopped meats (cutlet, etc.), or stewed meats (veal, chicken); with these may be taken potato purée or a little compote and a slice of wheat bread; at 5 o'clock, coffee, or, better still, a cup of wine soup, 1–2 eggs beaten up in white wine; at 8 o'clock, tea and a portion of roast meat, the same as at dinner, or a dish prepared from eggs with chopped ham; with this some bread. Tea may be substituted, in some cases, by a glass of white wine or beer. Modifications will have to be made in this bill of fare according to the disease; in the treatment of diabetes mellitus, adiposis and diseases of the liver, an entirely different régime must be instituted. In addition, I would like to observe that the use of the alcoholic beverages which irritate the intestinal canal must, if possible, not be allowed; on the other hand, a wine soup, given 2–3 times a day, may be ordered in certain cases; this is generally well borne by most patients suffering from gastric disease. The regulated diet prescribed must always be insisted upon, as the majority of patients who need a so-called Carlsbad treatment have acquired their disease by an irrational diet or excesses at table; therefore, much may be expected from a rational régime continued for some time.

SECTION 17.

USE OF THE SPRUDEL SALT IN WASHING OUT THE STOMACH—APPARATUS FOR GASTRIC IRRIGATION AND ASPIRATION.

For use in washing out the stomach, a solution is generally made of 2–3 tablespoonfuls of Sprudel Salt in 2–3 litres of river or rain water, heated to from 50°–55° C. The solution is generally used warm. It is not justifiable to undertake to wash out the stomach without first obtaining the assurance of the absence of contraindications (great atheromatous degeneration of the blood-vessels, advanced valvular disease of the heart, bleeding gastric ulcer or degenerating carcinoma of the stomach). It is also necessary to first accustom the patient to the continued retention of the stomach tube. This is absolutely necessary in a successful treatment. As to time of day, it will be found that the operation is best performed late in the evening, as at this time the stomach is emptied of contents that frequently irritate the gastric mucous membrane—a condition of things always present in gastric dilatation and mechanical insufficiency. If no supper is allowed after the operation, the diseased organ will be allowed more than 12 hours rest, and the full effect of the operation and of the Sprudel Salt upon the stomach is not altered by food. In addition to this, there is the agreeable assurance from the patient the next day that he has spent the night without distress—quietly sleeping—awaking in the morning with an appetite.

In the following diseased conditions I found the operation of washing out the stomach with solution of Sprudel Salt of service:—

1. In many cases of gastric symptoms associated with *loss of appetite and bad taste in the mouth*, but in which, by internal examination of the stomach, no severe functional disturbance could be discovered, the dyspeptic symptoms disappeared after washing out the stomach three times. The positive action in these gastric cases, which probably have a nervous origin, may, in all probability, be referred, in great part, to the psychical action of the manipulation.

2. *Chronic acid hypersecretion*, a very frequent form of disease, but which is by some authors described under the not very appropriate name of *dyspepsia acida*, as being particularly infrequent, may be looked upon as the first stage of catarrhal gastric disease. This is distinguished by the large percent-

age of hydrochloric acid in the gastric contents at all times of the day (even in the empty stomach), by motor disturbances of the stomach, as shown by the frequent finding of changed greenish or bluish bile-coloring matter in the stomach, as also by the much-delayed digestive mechanism, which appears to be dependent upon spasm of the pylorus, caused by the irritation of the gastric contents. The anatomical changes in the gastric mucous membrane in this form of disease may be demonstrated in the gastric contents by the excessively large quantity of sharply-defined, round-cell nuclei, which appear singly or in groups of 2-3, frequently glistening, and which are deeply stained by anilin. The origin of these must be ascribed to the glands of the fundus and to the destruction by the intensely digestive gastric juice of the glandular cells, of which nothing is left but the nuclei. For if, for instance, a piece of mucous membrane that has been torn from the fundus of the stomach by passing the stomach tube, is placed in a strongly acid, natural gastric juice, and the proper digestive temperature maintained, and examined daily under the microscope, the disappearance of the elements will be noticed every day more and more, with the appearance of numerous cell nuclei and fat globules. There is no better remedy for this diseased condition than washing out the stomach with warm solution of Sprudel Salt. The pretty concentrated (three to five per cent.) solution is applied in the evening. When there is great acidity of the gastric contents, when even after washing out the stomach in the evening an appreciable quantity of strongly acid liquid collects in the stomach in the morning, which is always the case when this condition is associated with gastric dilatation, it is of advantage to wash out the stomach with solution of Sprudel Salt early in the morning before any food is taken; *i. e.*, twice daily. A number of such cases I have treated here in Carlsbad with success by washing out the stomach with large quantities of Muehlbrunnen water, allowing no supper and ordering the proper dose of thermal water to be drank early in the morning, fasting.

3. The above described condition is generally found in stomachs affected with ulcer, especially of the pars pylorica. In these cases the washing out of the stomach with solution of Sprudel Salt produces the same beneficial effect as in the latter cases, by stopping the frequently hyper-acid secretion of the gastric juice, which must aid in the formation of the ulcers. If, however, there is well-founded suspicion of gastric ulcer, the washing out of the stomach must be made with much greater care than in the former case. The solution of Sprudel Salt must be more diluted (one to two per cent.), only used once daily—in the evening—for the first three days, then every other day, and later once or twice weekly. For in the same manner in which an excess of acid may injuriously affect the diseased places, so also may an excess of alkali prove deleterious.

4. In the second stage of gastric catarrh, in which the gastric acid secretion has given place to the *mucus secretion*, a condition of things so frequently found in drinkers, the washing out of the stomach with Sprudel Salt

solution affords symptomatic relief of the gastric ailments. This condition, which is diagnosed under the microscope by numerous chalice epithelial cells, granulated cutaneous cells filled with fatty dots, also isolated fat globules in the gastric contents, is, however, objectively not much improved by the use of solutions of Sprudel Salt in this manner, still the appetite is generally improved after its use, and the fullness and disagreeable oppression in the stomach are also removed. The stomach must be washed out for the first few days with more concentrated solutions (3 per cent. to 4 per cent.) so as to rid the walls of the adhering mucus and to stimulate the mucous membrane; the later treatments should be made with weaker solutions and at intervals of two or three days. For the latter cold solutions made with soda water are to be preferred.

5. The above cited condition approaches that found in carcinoma of the stomach, in which, according to my researches, a neutral or even alkaline gastric fluid is generated, which, however, possesses no digestive power even after being acidulated with HCl. The action of the washing of the stomach with solutions of Sprudel Salt in these cases is, however, much less than in the former. The patient only experiences relief in those cases in which the new formation has been the cause of copious accumulations of mucus or dilatation of the stomach with decomposed food remnants.

If it has been decided to wash out the stomach with solutions of Sprudel Salt so as to afford relief in these cases, it is necessary to use all the cautionary measures enumerated in cases of ulcer, under 3. It is generally found, in washing out carcinomatous stomachs, that, even if great care is exercised, very numerous pieces of mucous membrane, tinged with blood, will be brought to light, that are characteristic of this degeneration.

From the foregone it will be noticed that washing out the stomach with Sprudel Salt solutions may accomplish the following, either: (*a*) *Reduction of the excessively intense acid gastric secretion;* the result of the treatment to be determined from time to time by internal examinations of the stomach. (*b*) Ridding the organ of *irritating contents*, as microörganisms or products of digestion undergoing decomposition; information which is necessary to gain by examining the contents of the fasting stomach from time to time microscopically. (*c*) Symptomatic removal of *subjective gastric symptoms*, which is successfully accomplished even in many nervous cases. Or lastly, (*d*) The mechanical relief of the stomach by increasing its mechanical power, which, of course, is not peculiar to washing out the stomach with any particular solution. According to my opinion, this latter therapeutic factor is of most importance; for according to my observations, dating over many years, the delay of the passage of the contents of the stomach through the pylorus is in by far the majority of cases the real cause of the gastric ailment. Whilst great changes in the digestive chemism may be endured for a long time without subjective gastric ailments, these latter very speedily show themselves as soon as a change in the digestive mechanism supervenes.

In reference to the modus operandi of washing out the stomach, there are several methods that may be employed. In many cases the simple siphon contrivance* of Ziemssen, or the one modified by Rosenthal, may be employed. For ordinary cases this is the simplest and most convenient method. It, however, frequently disappoints the practitioner: a certain degree of rarefaction of the air in the stomach, a lurch with the stomach tube, interrupts the flow of the contents of the stomach; it becomes necessary, in order to reëstablish the flow, to pour in more water; concerning the quantity of liquid remaining in the stomach, however, there always remains some doubt. In cases of dilatation of the stomach, or when the same is filled with solid particles of food, such as are especially under consideration, the expedient generally fails. It is also impossible to make a more careful examination of the stomach by this method. It is necessary to have recourse to suction pumps. The ordinary suction apparatus with rod and piston, on account of the great suction power which cannot be regulated, is too dangerous to the mucous membrane, and is also unhandy.

The best apparatus, both for aspiration and irrigation, one which I have used for years for experimental, diagnostic and therapeutic purposes, is illustrated on page 81.

A is an aspirator bottle holding a number of litres, filled with ordinary water; at its lower part near the bottom is the tubulure g closed hermetically with a rubber stopper through which the small tube gh passes; to this is attached a rubber hose having a clamp at h and terminating in another bottle of the same construction D. The aspirator bottle is hermetically closed

---

* After going to press, Prof. Ewald published a procedure by which to obtain the contents of the stomach without the help of any aspirating contrivance whatever. (*Berl. kl. Wochenschr.*, 1886, Nr. 3.) The procedure consists in passing a soft stomach tube into the stomach, and then getting the patient to cough and strain with the diaphragm, and at the same time exerting pressure by means of the hand over the stomach, producing artificially vomiting and retching movements. This procedure may, in the absence of a stomach pump, assist the physician in an emergency, but can, in fact, only be looked upon as a makeshift; as the experiments instituted for the purpose by me have shown: 1. Frequently only so small an amount of fluid is obtained from the stomach that it will not suffice for all the tests; if, however, after this procedure, the stomach is aspirated, quite a large quantity of gastric fluid may still be obtained. 2. If the stomach contains solid particles of food, these will completely obstruct the openings in the stomach tube, so that nothing will pass out, unless the contents of the stomach are vomited and pass out by the side of the tube and are mixed with mucus from the mouth. 3. This method cannot be used to determine when digestion is ended, as all the particles of food cannot be obtained from the stomach. 4. When exact experiments are made it cannot be hoped to obtain accurate results, as it must be assumed that the forcible efforts at vomiting must alter the secretions and possibly cause the bile to flow into the stomach. 5. The intense efforts at retching and vomiting which are made during this procedure make the use of the same appear hazardous, when there is suspicion of ulcer, carcinoma of the stomach, in heart disease, severe emphysema of the lungs, atheroma of the blood-vessels and in general debility. In some cases, however, on account of its simplicity, this procedure may be recommended, for instance, when the ice-water method is used.

with a rubber stopper l, through which is inserted the small tube w. The bottles A and D may be advantageously replaced by similarly constructed tin vessels.

B is a vessel holding several litres of Sprudel Salt solution; i is a glass siphon tube. Instead of B any kind of vessel or bottle may be used.

C is a graduated bottle, or a similarly constructed vessel, filled with distilled water for experimental purposes. This vessel is supplied below at k either with a tube and discharging pipe l, or else with a siphon tube the same as in the vessel B.

E is a receiving bottle holding 2-3 litres hermetically closed with

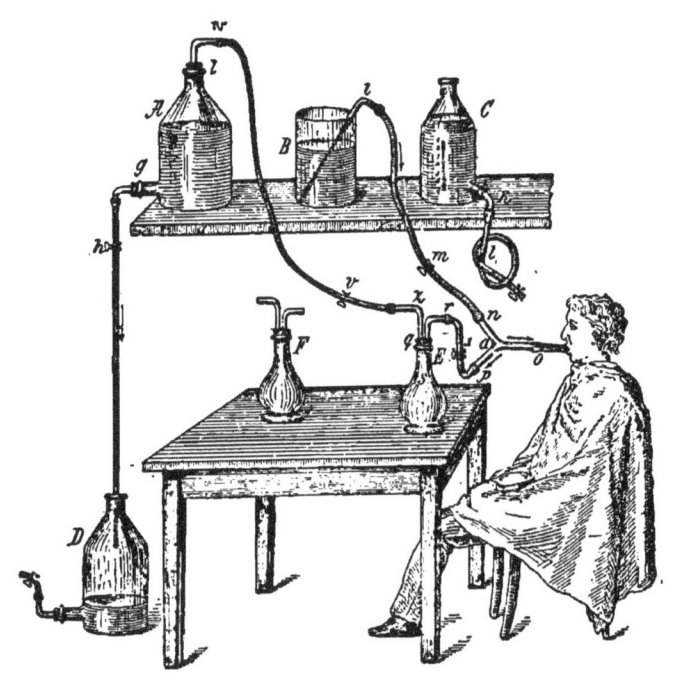

a rubber stopper q, having two openings; through one of these openings passes the shorter and narrower glass tube z for the purpose of connecting with the aspirator bottle A by means of the rubber hose $v z$, which may be closed at v; through the other opening passes the bent glass tube $s\ r$, which extends lower down, having a larger bore (1-2 ctm. in diameter). The stopper q may be very conveniently replaced by a stop-cock having a double bore. F is a reserve vessel constructed exactly like E. It is of advantage to have a number of such bottles in reserve.

n, o, p is a T-shaped glass tube having a diameter of 1-2 ctm. The one arm, $a\ o$, is inserted into a soft rubber stomach tube, the connection being made air-tight; if the latter is funnel-shaped at *o*, which is a convenient arrangement, a perforated rubber stopper must first be inserted into the funnel and then the arm *a o* inserted into the opening in the stopper. By means of the second arm, *a n*, connection is made at *n* between the Sprudel Salt solution, *B*, and the stomach tube *o*. This connection may be cut off by means of a clamp, *m*. The third arm, *p a*, connects the stomach tube *o* with the receiving bottle *E* by means of a rubber tube, *r p*, which may be cut off at *x*.

If, now, it is desired to wash out the stomach by means of the above-described aspirator, no regard need be had as to whether the stomach be empty or full. All that is necessary is to open all the clamps on the siphon tube *i*, *m*, *n*, and thereby allow it to fill up with the Sprudel Salt solution. When this has been done all the clamps are closed, excepting the one at *m*, through which the desired quantity of solution flows into the stomach through *n o*. Now the clamp at *m* is also closed, and time is allowed for the Sprudel Salt solution to act upon the gastric contents when the clamps *h*, *v*, *x* are opened (*m* remains closed) so as to aspirate the contents of the stomach. As regards the opening of the clamps, which must always be supplied with long handles, they are most conveniently pushed over on to the adjacent glass tubes. The bottle *A* now aspirates the air from *E* and the liquid out of the stomach, which flows in a stream through the tube *r s* into the receiving bottle *E*. When the liquid ceases to flow into *E* or the aspiration suddenly stops, *x* (and *h*) are closed (or compressed with the fingers) and *m* is opened, to allow of the flow of another quantity of Sprudel Salt solution into the stomach, whereupon *m* is also closed; time allowed for the action of the solution on the gastric mucous membrane, and then *x* (and *h*) are opened, whereby the gastric contents are transferred to *E*. If the air in the stomach has become greatly rarefied, the aspiration will be interrupted. It is then necessary, after closing the rubber tube at *x*, to squeeze the same obliquely at *p* with the fingers, so as to introduce the air into the stomach, which takes place with a hissing noise, when *x* is again opened. The stomach-tube dare not be moved to and fro in case the air has become rarefied, until air has first been introduced into the stomach in the above described manner at *p*, or else the gastric mucous membrane is pressed or sucked into the fenestra of the stomach-tube. If, by repeated introduction and aspiration of the saline solution from the stomach (a sort of stomach douche), the bottle *E* has become filled, the clamps *v* and *h* are closed and the bottle exchanged for the reserve bottle *F*. In the same way, as soon as the bottle *A* has become empty and *D* filled, *A* is placed upon the floor and *D* in its place, with the stopper *l* closed hermetically, to act as aspirator bottle. It is more convenient for the experimenter to place the aspirator bottle *A* by the side of the patient, beside the bottle *C*. The physician who watches the patient and who must hold the stomach tube at *o* with his fingers, can thereby regulate and if necessary stop the flow of water, and so prevent any waste. This apparatus will be found to be very convenient when the operation of washing out the stomach must be done frequently, and repeated observations and experiments made. One can thereby operate with the greatest convenience and assurance, and without interruption, arriving at the most accurate experimental results in a manner the least disagreeable to the patient. The apparatus may be constructed by the physician himself in any hospital and modified to suit the convenience and exigencies of the case.

SECTION 18.

USE OF THE SPRUDEL SALT PER RECTUM.

The Sprudel Salt is used by the rectum in a warm solution (36°–40° C.), preferably made with rain or river water. A solution of 2–3 tablespoonfuls of Sprudel Salt in 2 litres of water is preferred. The solution is best introduced by means of Hegar's funnel, in the evening, before bed-time; it is only to be done in the morning in case it disturbs the sleep. The injections may be either large and abundant or small. The above solution is used for the *abundant or copious injections;* it is necessary in this case to try to introduce enough solution so that part of it may get into the cæcum. To accomplish this 2–3 litres are necessary, and when the solution reaches the cæcum the patient experiences in it, as also in the entire abdomen, a feeling of tension. The most judicious manner in which to order the injections is the following: Enough solution is introduced into the rectum until the patient experiences a strong feeling of tension; the patient is then to assume the horizontal position, lying on the back, and remain so (about ten minutes) until a strong desire is experienced to go to stool, which the patient is to give way to, whereupon the patient shall continue evacuating the bowels as long as anything will come away. The patient may be recommended to assist the evacuations by stroking the abdomen with the hand in the direction of the large intestine, beginning at the cæcum, and at the same time making efforts at straining. Should there continue to be a desire to go to stool after the main passage the patient is to give way to the same, and should this be the case during the night he should get up and have a passage, as, should he fail to do so, the fecal masses forced into the large intestine from above would continue to fill it up for a long time. In some patients the large injections cannot be used, as they are not able to retain such large quantities of liquid; hence, it is difficult in these cases to clean out the large intestine and to act locally upon it. One must be satisfied to use small injections in these cases; these are also better borne by patients with heart disease.

Small injections are made by dissolving 1–2 teaspoonfuls of Sprudel Salt in a teacupful of either river or rain water, introduced into the rectum preferably in the evening, and retained as long as possible. If the small injections have for their object the introduction of the salt into the circulation of the blood-vessels, they must be retained all night, in spite of the desire to go to stool. If they are employed to evacuate the bowels, the desire to go to stool may be heeded after 1-2 hours. When the Sprudel Salt is given by the rectum for a long time, capillary hemorrhages occasionally set in. In that case, the injections must be stopped for several days, or less concentrated solution or even pure water are to be used, or, finally, as will be found very judicious, give the injections every other night.

The Sprudel Salt may be used in the form of injections in all cases with the exception of the contra-indication cited in Section 13, and deserves much more attention than heretofore. It frequently happens that injections of plain water used to evacuate the bowels do not produce any discharge of fecal matter; the addition of Sprudel Salt in these cases will accomplish the desired result. In intestinal therapeusis the gastric function is saved by the injections, in addition to which they act locally upon the part affected. In diseases of other abdominal organs, the injections do not only save the gastric function, but enable larger quantities of the Sprudel Salt to reach the circulation of these organs. It is therefore always best in certain cases to divide the dose of Sprudel Salt so that one part will reach the circulation per os, the other per rectum.

SECTION 19.

LENGTH OF TIME AND DIRECTING OF THE SPRUDEL SALT TREATMENT.

The length of time of the treatment naturally depends in every case upon the success; a term that would be applicable to all cases cannot be given. Another question in this direction may be of importance: how long the Sprudel Salt may be taken without detriment to the organism. The effect of the continued use of the Sprudel Salt shows itself mainly in the change of the gastro-intestinal function and in the general health. According to my observations, I would infer that the continued use during six weeks of small and medium doses (5-10 grm.), and during four weeks of large doses (15 grm.) could be borne under ordinary circumstances without great alteration of the organism. If, in a special case of disease, success follows the use of the Sprudel Salt treatment, but it is noticed that the treatment will require longer than four or six weeks, it is well to divide the same into periods of three to four weeks, allowing an interval of from one to two weeks to intervene between.

It is absolutely necessary to direct the treatment, as thereby it is not only possible to give assurance of the progress of the treatment, but also data to determine the dose of the Sprudel Salt, the length of time the treatment is to last, if the same is to be continued, interrupted, or considered as ended.

The direction must, however, be based upon objective clinical examinations: In diseases of the liver upon physical examination of the liver; in diseases of the urinary apparatus, in diabetes, upon chemical analysis of the urine; in diseases of the intestines upon examination of the stools; in adiposis by determining the body weight, but in this case the degree of muscular weakness, and above all the action of the heart and the possibly engendered intestinal catarrh, must be taken as a guide, whether it will not be well to discontinue the treatment for a time. Especially in diseases of the stomach is it necessary to control or direct the treatment; hence I will here make a few observations in this connection from my own experience.

Just as it is necessary to determine, by internal examination of the stomach, the degree of acid secretion, power of digestion and gastric capacity, if the Sprudel Salt treatment be indicated and what dose is to be employed, so also is it necessary to direct and control the treatment, and the most proper time to make the internal examination is the third week of the treatment. If, for instance, it will be found that in acid hypersecretion the acidity of the gastric contents has only been reduced very little, it will then be necessary to increase the dose of the Sprudel Salt, but if a very great reduction is found it will be necessary to reduce the quantity of the salt. If the degree of acidity, determined by the modified ice-water method, be six degrees, and just as much after an albumen digestion lasting a half to three-quarters of an hour, then the treatment may be considered as ended, as a further continuance of the Sprudel Salt would produce a total acid insufficiency and digestive incapacity. This cannot be determined from the subjective symptoms of a patient with a gastric disease, and, therefore, no positive deductions as to the favorable or unfavorable course or the termination of the treatment can be drawn therefrom. It is, for instance, impossible to assert positively, judging by the subjective symptoms, if the gastric contents are strongly acid or entirely devoid of acidity, inasmuch as the most diverse objective pathological conditions are described by patients just alike. One thing I have often noticed in treatment with Carlsbad Sprudel Salt or Carlsbad water, when, before the treatment, a patient suffering from excessive hypersecretion complained of a voracious appetite, this would, after a certain time, be much diminished and give way to a sense of satiety. This occurred in the period when the acidity was much lowered. If, however, a case of acid insufficiency is treated with small doses of Carlsbad water or Sprudel Salt, then the first thing to be determined is whether it is a case in which the normal secretion may be lowered, or the anatomical change in the secretory apparatus is such as would appear irreparable. If the internal examination, after two weeks of treatment, shows that the original acidity is not at all increased or has even become less and the secretion of albumen from the stomach is delayed, then success from a continuation of the treatment is not to be expected; there is even danger of destroying the remaining digestive chemism by the continued use of the Sprudel Salt. Subjective sensations are here also not much to be trusted. In some favorable cases of gastric insufficiency treated with Carlsbad water I have noticed that during the course of treatment the patients asserted that they got a good taste in the mouth, had a very good appetite and even felt a sort of hunger. In these cases the internal examination showed an increase in the originally very low degree of acidity of the gastric juice. It may again be mentioned that there are very much fewer cases of acid insufficiency than of hypersecretion, and that in the former cases there are fewer cases benefited.

It may be easily seen from the foregoing that it is necessary, in every treatment of a gastric disease with objective success, to examine the stomach internally before and during the treatment, and that the cessation of the subjective symptoms cannot be taken as a criterion of the cure of the disease. In what manner to examine and direct a gastric treatment may be gleaned from the cases in Section 11. It is, however, seldom, especially in Carlsbad, that an objective control of the treatment is possible. It becomes necessary to be satisfied with misleading subjective symptoms. When, for instance, in a gastric affection in which the previously voracious appetite disappears, or when the previous want of appetite is improved by a Carlsbad treatment, and gives place to a loss of appetite and gastric distress, it then becomes necessary to discontinue the treatment.

CONCLUDING REMARKS.

To what constituent of the Sprudel Salt may especially be ascribed the physiological and therapeutic action produced, cannot positively be affirmed, because the action of the numerous components of the Sprudel Salt have not yet been clinically investigated upon man. The beneficial action of the Sprudel Salt is rather to be found in the condition which Prof. Leube * emphasizes for the Carlsbad water, which "amongst all other alkaline saline springs has taken the first place in gastric therapeutics," namely, in *the happy combination of its constituents*. The Sprudel Salt may well be compared to " an extract of the Carlsbad water," possessing the happiest combination of the most active saline remedies which are present in the Thermal water. Besides this there is a second important factor to be considered, that the many saline substances present in the Sprudel Salt, and demonstrated by chemical analysis to be present there in only very small quantities, produce a very decided action upon the intricate actions of the cell elements of the gastro-intestinal mucous membrane ; and it is on this account that the original natural Sprudel Salt will always be superior to the artificial product.

---

* Ziemssen. "Handbuch der speziellen Pathologie und Therapie." VII, H. 2, pp. 109, 110.

APPENDIX.

SECTION 20.

THE RELATION OF THE SPRUDEL SALT TO THE CARLSBAD THERMAL WATERS.

As the Sprudel Salt had its origin in the desire to furnish a product containing all of the constituents of the Carlsbad waters, my desire has been to ascertain if the pharmacodynamic and therapeutic action of the Sprudel Salt and Carlsbad water are identical or not, inasmuch as from a practical standpoint it was desirable to ascertain whether or no, clinically, the Sprudel Salt might be used instead of the Carlsbad waters. At first nothing seemed easier than to make parallel experiments upon the same individual to ascertain the influence upon the gastric function once with the Thermal waters then with the Sprudel Salt. The insurmountable difficulty, however, arose of placing both products under the same experimental conditions. As is well known, the Carlsbad Sprudel Salt is produced by evaporation of the Carlsbad Sprudel water and the removal of the insoluble constituents. One litre of Sprudel water, according to Ludwig's and Mauthner's analysis, produces 5.5168 grm. dry residue in which there are 4.9527 grm. soluble constituents. It may, therefore, be assumed that this latter quantity of Sprudel Salt dissolved in one litre of distilled water would be equal to one litre of Sprudel water as regards the soluble constituents. When, however, I dissolved 4.9527 grm. of Sprudel Salt in distilled water and determined in the artificial water thus obtained the main constituents, sodium carbonate, sodium sulphate and sodium chloride, quantitatively, I found that it did not entirely correspond to the neutral Sprudel water as far as these constituents were concerned. I found by analyzing, with the same reagents and at the same time, the bottled Sprudel water and the imitated Sprudel to contain in 100 parts of liquid the following :—

	Bottled Sprudel Water.	Sprudel Water made from Sprudel Salt.
Alkalinity,	35.4 c. c.	22.4 c. c. $\frac{1}{10}$ normal sulphuric acid.
Chlorides,	22 6 c. c.	20.4 c. c. $\frac{1}{10}$ normal silver-nitrate solution.
Barium sulphate,	0.494 grm.	0.381 grm.

From the above it will be seen that the natural Sprudel water is much stronger so far as the main constituents are concerned, and what concerns us most here, the relation of these constituents one to the other has been changed; although the percentage of chlorides in both liquids did not differ much,

APPENDIX.

the percentage of sulphates was diminished by $\frac{1}{5}$, and the alkaline carbonates, the most important agent upon the gastric function, to almost ⅓ in the Sprudel Salt solution. The influence of these liquids upon the organism cannot, therefore, be compared in figures. It is impossible to find a quantity of Sprudel Salt which, when dissolved in water, would be quantitatively of equal value with the Sprudel water. From a chemical standpoint of view there are two quantitatively heterogeneous solutions under consideration.

1. Still, I have made parallel examinations with the above mentioned Sprudel Salt solution which was intended to imitate the Sprudel water and the Sprudel water itself, to ascertain the action of both upon the gastric function. Case XXI, 114, 115, of the table, is one of these cases. It was found, as might be expected, and was observed in a number of other cases, that the depressing or lowering effect of the Sprudel Salt solution upon the stomach was greater than that of the natural Sprudel water. The acidity appeared higher after taking the Sprudel water, although it contained more alkaline carbonates, the sulphates disappeared more rapidly, and the gastric juice appeared more capable of digesting than when the Sprudel Salt solution had been taken.

2. As it did not appear opportune to test both remedies pharmacodynamically upon the ground of chemical principles, the question was taken up from a practical clinical standpoint. The pharmacodynamic action of the usual minimal doses of Sprudel water and Sprudel Salt were compared. For this purpose, with accurately weighed and measured quantities of 250 c.c. of Mühlbrunnenwasser on the one hand, and 5 grm. Sprudel Salt in 250 c.c. distilled water on the other, parallel experiments were made on the same individuals under the same conditions as regards gastric function. The mode of examination was identically the same as given above in Section 2. The experiments made for this purpose are mentioned in the table in their proper place. The experiments not only demonstrate that, after the ingestion of a quarter litre of Carlsbad water, a gastric juice of acid reaction, and capable of digestion, appears sooner than when 5 grm. Sprudel Salt has been taken, which quantity really represents a litre of the Thermal water, but that the maximum of acidity after the Carlsbad water is much higher in less than half an hour than after the Sprudel Salt; it is even higher than the acidity obtained by the irritation of ice water. In case I, 5, the acidity with Sprudel Salt after half an hour was 3.6, with Mühlbrunnenwasser, however, 14.0; in case V, 28, with Sprudel Salt, 1.6; with Mühlbrunnenwasser, however, 13.2; in case IX, 46, with Sprudel Salt 3.4 degrees alkalinity, with Thermal water 12.4 degrees acidity. In case XIX, the figures were 3.4 and 22.0; in case XXV, 0.8 and 15.4. In alkaline gastric juice the alkalinity after Carlsbad water was less than after Sprudel Salt solution with the sam mode of experimentation; in case X, 57, degree of alkalinity, after Spru Salt solution, 2.0, and after the Thermal water, 1.2. In case XIII, 7

differences were 4.0 and 2.8. In these experiments, therefore, the super-irritated gastric function is relieved *more by the Sprudel Salt* than by the Thermal water, which is very probably owing to the larger quantity of fixed constituents of the Sprudel Salt.

3. If the facts set forth under 1 and 2 are placed in juxtaposition, and the results which I have obtained and published concerning Carlsbad water compared from a pharmacodynamic point of view with the results of my experiments with Sprudel Salt, the following difference will be noticed between the Carlsbad water and the Sprudel Salt from a practical point of view.

(*a*) Single doses of Carlsbad water stimulate the acid secretion of the gastric mucous membrane to a greater degree than single doses of Sprudel Salt. The maximum of acidity after Carlsbad is not only much higher, but is reached much quicker than after Sprudel Salt.

(*b*) The gastric juice becomes capable of peptonization sooner after Carlsbad water than after Sprudel Salt.

(*c*) The stimulation to acid secretion lasts much longer with Carlsbad water than with Sprudel Salt, so that I had to wait hours in my experiments with Carlsbad water until the degree of acidity was reduced to that of the empty stomach.

(*d*) The salts disappear more rapidly from the stomach after the ingestion of the Carlsbad water than after taking a Sprudel Salt solution.

(*e*) Warm Carlsbad water is more stimulating to the gastric function than cold, whilst with Sprudel Salt the opposite is the case.

(*f*) Whilst the quantity of Carlsbad water ordinarily taken influences the intestinal function but very little, the ordinary doses of Sprudel Salt exert a very marked influence upon the entire intestinal canal.

(*g*) As regards the final result of a continued use of Carlsbad water and of Sprudel Salt, both remedies correspond very closely with one another. Given in large doses the digestive chemism is lowered by both, and frequently stimulated by small ones. The general health and nutrition is affected in the same way by large doses. The alteration in the intestinal function may be noticed more frequently and in a higher degree after Sprudel Salt than after Carlsbad water.

4. As regards the clinical deductions derived from these observations, the following may be assumed therapeutically:—

(*a*) Carlsbad water is to be preferred in many cases of independent gastric diseases, whilst Sprudel Salt is especially indicated when the stomach and intestinal canal are both involved, but also is of most service in dilatation of the stomach associated with acid hypersecretion, as small doses appear well suited to lower the gastric function.

(*b*) Sprudel Salt has qualitatively the same therapeutic effect as the Carlsbad water, but quantitatively, at least in one respect, not: Carlsbad water is more stimulating.

(*c*) Whether or no the Sprudel Salt is to be preferred to the Carlsbad water in other diseases I am not prepared to say, on account of the want of comparative study; speculative deductions would be of no use, and in practice even dangerous, because it is impossible to know if the changes in the more distant organs are produced in the organs themselves by the direct action of the solution of Sprudel Salt introduced, or whether they are the result of reflex action.

SECTION 21.

THE ACTION OF THE THERMAL WATER AND SPRUDEL SALT WHEN USED JOINTLY.

I have studied the Sprudel Salt in still another connection with the Carlsbad water. It is customary in Carlsbad to increase the action of the Thermal waters by the addition of 5 grm. Sprudel Salt to the glass of water, to avoid the drinking of too large quantities of water at one time; or rather, it is ordered by the physicians here, to act upon the bowels. It has been studied how a solution of 5 grm. Sprudel Salt in 250 c.c. Mühlbrunnen or Sprudel water behaves in the stomach. The experiments made to determine this, and cited in the table, were upon the individuals XVII, XVIII, XIX, XXII; they show in every case that after taking such a mixture, compared with Carlsbad water alone, the alkalinity of the gastric contents lasts comparatively very long, the sulphates remain a much longer time in the stomach, the acid secretion appears much later, the maximum of acidity is much less, and the digestive power of the gastric juice is much diminished. In case XVII, 87, 89, after taking Mühlbrunnen water alone, a gastric juice capable of digestion with an acidity 3.6 appeared after 2 quarter-hours; after taking the mixture, a gastric juice appeared incapable of digestion with an alkalinity 22.0. In case XIX, 98, 99, the acidity after 1 hour was respectively 22.0 and 3.4, and not until 2 hours after taking the mixture was an acidity of 8.4 reached. In case XX the reaction 2 quarter-hours after taking the Mühlbrunnen water was 10.6 acidity; after taking the mixture, 22.8 alkalinity and possessing no digestive power. After repeated doses of the Thermal water alone, on the one hand, and of the mixture on the other, the relation of the acidity was as follows: In case XXIII, 93, 94, after 5 quarter-hours, 5.6 acidity, but 12.5 alkalinity, when repeated doses of the mixture were given. If the behavior of the mixture taken by itself is compared with a solution of Sprudel Salt in distilled water, the same differences qualitatively will be noticed as when the Thermal waters and the mixture are compared, only the differences are quantitatively less. In case XVII, 84, 89, the acidity after Sprudel Salt is 1.0; after taking the mixture the alkalinity is 22.0. In case XIX, 101, 102, after 2 hours the acidity is respectively 16 and 8. In case XIX, 103, 104, the acidity is 5.0 in 5 quarter-hours after taking 2 doses of Sprudel Salt; after two doses of the mixture the alkalinity is 30, and a precipitate in the test for sulphates. It follows, therefore, that mixtures of Thermal water and Sprudel Salt remain much longer in the stomach than when either are taken separately, that the acid secretion and digestive power are affected much longer by the mixture than when Carlsbad water or Sprudel Salt alone is administered; that the maximum of acidity, under the influence of the mixture, not only appears later, but is less, than when either alone is taken. Taken as a whole, the mixture of Sprudel Salt and Carlsbad water reduces the digestive chemism in a high degree, wherefrom the following practical deductions may be made:—

(*a*) The mixture of Thermal water and Sprudel Salt is of advantage in strongly-acid hypersecretion, especially when complicated with dilatation of the stomach, and perhaps by disease of the biliary passages. In cases of gastric acid insufficiency it is positively to be avoided.

(*b*) The intervals between the doses of the mixture must not only be longer than when the Thermal water alone is taken, but also than when Sprudel Salt alone is employed; they must be set at 2–3 quarter-hours. Breakfast must, accordingly, be postponed much longer after the use of such a combination, and must not be allowed before the lapse of two hours, if it is at all desirable to have some regard for the gastro-intestinal function and to get the full benefit of the food.

(*c*) Only the first doses of Thermal water are to be taken with the addition of Sprudel Salt, the following doses to consist of Thermal water alone, and not *vice versâ*, because in the latter case concentrated solutions come in contact with the stomach already debilitated by drinking water, affect its function still more and remain in the stomach too long, so that they enter the intestine very late, and therefore also generally fail to move the bowels. To make more than two additions of Sprudel Salt of 5 grm. to a glassful of Thermal water would tax the gastric function too much. If, therefore, 3 glasses of Thermal water be ordered, with two additions of Sprudel Salt, 5 grm. are put in the first glass; in half an hour the second is ordered, with the second 5 grm. After 3 quarter-hours the third glassful of Thermal water is drank alone, and not until the end of the second hour, or even later, after the last glass is breakfast to be taken. In adiposis, of course, the interval between the taking of the water and food may be reduced one-half.

SECTION 22.

EXPERIMENTS TO DETERMINE THE INFLUENCE OF EXERCISE UPON THE BEHAVIOR OF THE THERMAL WATER AND SPRUDEL SALT SOLUTION IN THE STOMACH.

Lastly, I have tried to elucidate another question experimentally. Does the customary promenade taken during the ingestion of the Carlsbad water, the Sprudel Salt solution, as also the mixture of the same, exert any influence upon their behavior in the stomach. I have had to disregard this factor from my previous examinations of Carlsbad water, as they were made during the winter months. As I have made my experiments with Sprudel Salt for the most part during the summer months of the year 1884, I embraced this opportunity to get the patients at one time to walk in the clinical garden (according to the suggestion of my colleague, Gluzinski) immediately after taking the solution, another time to remain quietly seated in the room. In these experiments, in addition to all the usual chemical examinations of the gastric contents made heretofore, the chlorides in the filtrate were determined quantitatively with one-tenth normal silver solution, and that for the purpose of obtaining more data to judge of the behavior of the saline constituents of the Thermal water and of the Sprudel Salt in the stomach. In the six cases examined and incorporated in the table of experiments it was found that in the case of Sprudel, as well as for Mühlbrunnen water (case XVII, 85, 86), the chlorides and sulphates disappeared from the stomach sooner after exercise than when the patient remained quietly seated. When a solution of Sprudel Salt, however, was used (experiments 84, 85) the reverse was the case, and when a mixture of Thermal water and Sprudel Salt was used (experiments 88, 89) the exercise as well as the resting remained without influence upon the behavior of the same in the stomach. In the case XVIII, experiments 91–95, it was found in all three combinations that the behavior of the same in the stomach was exactly the same when the patient remained quietly seated as when he walked about, although the aspirations were not made until after 5 quarter-hours of industrious walking. In case XX, 108–111, when Mühlbrunnen water was used, the conditions in the stomach were exactly identical after sitting and walking, whilst after taking the saline solution the alkalinity and the percentage of chlorides in the gastric contents were less after walking than after sitting. In case XXII, 117–119, where three doses of Sprudel water had been given, and aspiration not performed until after six quarter-hours, the quantity of the salts in the gastric contents were found to be less after walking, the acidity on the other hand diminished. The cases XXIII and XIX also failed to give a decisive result. These experiments therefore demonstrate :—

(*a*) That in cases in which bodily exercise manifested any influence, the same showed itself in the more rapid disappearance of the salts from the stomach, so that it would appear as if the disappearance from the stomach was hastened by the possible gastric movements. Walking, however, appears to exert no influence upon the secretion of acid.

(b) Whilst, therefore, walking is of decided advantage when drinking the mineral waters or solution of Sprudel Salt, to stimulate the intestinal function, promenading to influence the gastric function is not necessary, still, however, when possible, to be recommended. As it is generally desirable to influence the entire gastro-intestinal tract when mineral spring waters or Sprudel Salt solution are ordered, bodily exercise during a mineral water treatment is in place, as it offers the additional advantage of the general stimulating effect of a sojourn in the cool morning air.

In closing this treatise I feel constrained to tender my sincerest thanks to the Director of the Clinical Institute in Cracow, Professor Korczynski, who has also in this work kindly favored me with his very valuable scientific and material help.

DIETARY.

This dietary, which is meant to supplement the one in the body of the book, is by no means intended to be complete or as covering anything more than a few of the most salient points, still, it may perhaps prove of value in the absence of a more exhaustive work on the subject.

Briefly as the subject is treated here, it may not be out of place to call attention to a few cardinal principles relating to the taking of food and its digestion or solution.

A point to be remembered is that not so much depends upon the kind of food taken, as is generally supposed, always providing that it is sound and properly prepared, as upon the manner or way in which the food is eaten, the meals arranged, and the circumstances under which they are taken.

Let it be remembered that *digestion begins in the mouth*, with the very important acts of mastication and insalivation, here being the only place where the food can be comminuted and subdivided so that the saliva, gastric and other digestive juices may act properly upon it. If this one act is not properly performed, all the others will be more or less hampered, retarded, and in some cases entirely prevented from doing their share in the proper digestion or solution of the food.

If the meals be taken hurriedly, especially when heated and exhausted, discomfort will inevitably follow. If there is not time sufficient to take the food properly, at ease and in comfort, it is best not to take the meal at all at that time, but to substitute a very light lunch consisting chiefly of liquid food that is easily digested, and arranging for the meal proper when more time may be taken. The chief meal of the day, the dinner, may thus be postponed until after business hours and a light lunch substituted.

Such light lunch may consist of a cupful of bouillon, which is made by beating up the yolk of an egg into a cupful of hot broth or consommé, or a cupful of beef-tea may be taken with a piece of toast or cracker, or a glassful of Johann Hoff's Extract of Malt (Eisner), or even a glassful of plain hot water, if none of these be accessible; this latter, in most cases, is to be preferred to tea or coffee. Under no circumstances should there be large quantities of these liquids taken, and cold drinks ought never to be allowed.

We will now proceed to the consideration of the diet best adapted to patients suffering from diseases for which the Carlsbad Sprudel Salt (powder form) is especially recommended.

Acute Gastric Catarrh.—The indications with regard to food in this disease are to limit it in quantity and quality, so as to prevent any undue irritation by the food, and to allow the irritated and inflamed mucous membrane of the stomach as much rest as possible. Therefore nothing in the form of food proper must be allowed between meals. In severe cases food had best be abstained from altogether for a day, or even two days, nothing but Johann Hoff's Extract of Malt (Eisner) or hot water being allowed. In all these cases the food had better be liquid as much as possible, consisting of meat broths, milk diluted with soda water or lime water, whey, junket, etc. Solid food had better be returned to very gradually, and to assist its digestion be followed by Johann Hoff's Extract of Malt (Eisner). On recovery care must always be had to avoid any and every exciting cause.

Chronic Gastric Catarrh.—This often being a sequel to the acute, what has been said in reference to that also applies here. Always remembering to avoid the cause, be it indiscretions as to eating or drinking or over haste, or inopportune times in taking the meals. The dietetic treatment in these cases must be strict and systematic. In many cases an exclusive milk diet answers very well, and with the aid of the Carlsbad Sprudel Salt will frequently effect a cure. But in employing this diet it will always be well to have regard to the peculiarities and idiosyncrasies of the patient as regards the digestion of milk. Should there be any discomfort experienced, before abandoning this diet, it may be advisable to try giving the milk mixed with Giesshübler, a most delightful, natural aerated water, or with lime-water; in other cases it may be necessary to skim off the cream or add a little salt; again, in others, it may be well to peptonize or predigest it. No matter how the milk be taken it is always necessary to drink it very slowly, as when large quantities enter the stomach at once they may form one solid mass of curd that will be almost impervious to the gastric juice, and really act like a foreign body. Bearing this in mind it will, in some cases, be found advantageous to substitute buttermilk which, if fresh, will be better borne by the majority of patients. Niemeyer suggests that "when the patient is hungry let him eat buttermilk, and when he is thirsty let him drink buttermilk." In some cases it is well to employ the whole milk, skimmed milk and buttermilk interchangeably, and at times add to the milk either salt, bicarbonate of sodium, lime-water, or, best of all, some aerated water. * Another expedient that will be found to work well, when the milk is not well borne, is to either whip it or shake it thoroughly; this will make it seem much lighter, and it will be found easier to digest. To accom-

---

* The best of all aerated waters will be found to be those that are charged *by nature* with carbonic acid gas; these are far superior to any and all those which are charged artificially; they are better borne by the stomach, and do not cause the unpleasant distention and discomfort so frequently experienced when the artificial waters are used. Of all the natural aerated waters the Giesshübler will be found to answer the requirements in these cases best.

plish this the milk may be placed in an ordinary preserve jar, closing it and then shaking for a minute or two vigorously; of course, the jar must not be filled and not shaken too long. In the chronic gastric catarrh caused by the abuse of alcoholic beverages, it will be found best to mix the milk with carbonic acid water or plain soda, or better still, "Giesshübler" water.

As to the quantity to be taken, this must necessarily vary with the individual as well as with the severity of the case, varying from a few ounces every two hours at first, in severe cases, to a little over a pint taken three or four times a day. The majority of cases met with in practice will be found to be such as will improve very much under a plain bread and milk or crackers and milk diet.

The milk diet, however, will not be well borne by some patients, no matter how it is varied and how combined, and for those it will be necessary to get up a dietary that will in blandness and unirritating qualities approach this as near as possible. In these cases it is necessary to avoid starchy food as much as possible, and to give the patient nourishing broths, into which the yolk of egg has been beaten, with toast. Later on, when the severe symptoms have somewhat subsided, oysters, raw or panned, may be added to the dietary. Then fresh fish of the kind that is not too greasy; later on, the meat of fowl, game, etc. Always avoiding greasy and starchy food. The vegetables may be added in the following order, beginning with lettuce, then spinach, salsify, tomatoes, etc.; and not until the patient is almost well may he return to potatoes and pastry. Condiments and sauces are to be avoided. Coffee and tea as well as whiskey, brandy, beer, wine, etc., unless in very debilitated conditions, must not be allowed. In these cases it will be found that every indication will be met by Johann Hoff's Extract of Malt (Eisner).

Ulcer of the Stomach.—What has been said concerning chronic gastritis holds good in ulcer of the stomach, always bearing in mind that the object here is to give as little food as possible, and so allow of the healing of the ulcer. In cases where dangerous hemorrhages have occurred, it is best to keep the stomach absolutely quiet and feed by the rectum.

Dilatation of the Stomach.—Here the object is to avoid overloading the stomach by either solid or liquid food, hence to give the food in as concentrated a form as possible. Starchy and sweet food had better be avoided, and the régime laid down under chronic gastritis followed as near as possible.

Dyspepsia or Indigestion.—Here it will be necessary to inquire into the quantity as well as quality of the food, as to the manner and circumstances and time of taking it, etc. As to the kind of diet best suited to the case, it will be found that in some a plain bread and milk or crackers and milk diet will answer the purpose where there is no individual peculiarity or idiosyncrasy regarding the digestion of milk. In very severe cases the diet recommended for chronic gastritis will be found the best. In milder cases, the avoidance of the particular food that is known to be especially prone to bring on an attack, with a temporary abstinence regarding greasy food and pastry; in men, the cutting down of the allowance of tobacco and abstinence from drinking spirits and beer or wine will suffice, especially when the treatment with Carlsbad Sprudel Salt is carefully carried out.

The dyspeptic should be especially cautioned against eating immediately after severe mental or physical work, or performing such toil immediately after a meal. As to the use of tea and coffee, it will be found that in themselves they are in the majority of cases not hurtful; it is only when taken too often or in too large a quantity that they are harmful.

Chronic Constipation.—Here again the cause of the trouble must be inquired into, and if possible removed. The habits of the patient as to exercise must be looked after, etc. As a too concentrated diet, consisting chiefly of nitrogenous and too easily digested animal food, with very little or no residue to excite and stimulate the nerves of the intestinal mucous membrane, may be a cause, this must also be looked after. Then, again, the constipation may be caused by the too often repeated contact of overstimulating diet and too stimulating residue, thereby exhausting the excitability of the intestinal nerves; in other cases, too dry a diet may be the cause of the trouble, etc. Persons who live on a too exclusive diet of any kind must be directed to add a larger number and variety of dishes to their bill of fare. If green vegetables and fruits have been avoided, these must be added, etc. The various kinds of brown bread, those made from whole meal, may sometimes be advantageously added to the diet. When these or other foods having a large percentage of irritating, undigested residue have been used in excess, they had better be dropped entirely, as they in some cases will produce the very evil they are intended to correct.

Diarrhœa.—When uncomplicated with other disease is either acute or chronic, the one temporary, the other sometimes very difficult to cure. The rule in both cases to observe is to avoid all food that leaves a large quantity of undigested residue behind to irritate the mucous membrane. Green vegetables, acid fruits, nuts, potatoes, coarse brown bread, rich, greasy or acid dishes should be avoided.

The best diet in these cases is milk, which is best taken boiled or in the form of buttermilk, especially when the latter is fresh and obtained from sweet milk; should this not answer, Koumiss may be used instead. Next to milk in efficiency comes raw meat, then peptonized foods, and if these, for any reason, cannot be taken, then bland farinaceous foods, such as arrowroot, rice, tapioca, etc., flavored with cinnamon, nutmeg or cloves, may be substituted.

Diseases of the Liver.—These cases do best on a diet as free from fat and starchy food as possible. Coffee, beer, wine and spirits, as well as tobacco, are best avoided.

A mixed diet, consisting of lean beef and mutton, or chicken, with greens, acid fruits, especially lemons, is generally well borne. In very severe cases the diet laid down under chronic gastritis will be found of advantage.

Diabetes Mellitus.—Of this disease there may be said to be three different kinds of cases: slight cases, which, when Carlsbad Sprudel Salt is used, need no special diet; medium cases, which had better, in addition to the use of Sprudel Salt, submit to a somewhat restricted diet, and severe cases, which must adhere strictly to the diet prescribed. In these cases all articles of food containing starch or sugar must be avoided. Patients, however, may eat meats of all kinds (excepting liver), poultry, game, fish, oysters, eggs, cheese, butter, milk and cream. Of vegetables they may eat tomatoes, lettuce, celery, water-cress, chickory, dandelion, spinach; greens of any kind, such as turnip or beet tops, onions, cabbage, cauliflower, string beans, radishes, mushrooms, truffles, Brussels sprouts, asparagus, salsify, artichokes, cucumbers and pickles. All kinds of nuts excepting chestnuts. Of beverages they may have tea and coffee without sugar, Rhine wine, brandy, whiskey, gin, rum, arrack, apple-jack, the various aerated mineral waters, etc.

To be avoided is sugar and all food containing it, syrup, honey, bread, wheat and rye flour, cornmeal, arrowroot, sago, tapioca, oatmeal, barley, potatoes, beets, parsnips, carrots, rhubarb, peas and beans, sweet fruit, chocolate, sweet cider, malt liquors, champagne, sweet wines, etc.

As to the matter of bread in diabetes, this problem has always been the one most difficult of solution. All the diabetic flours and breads so far introduced have invariably contained a very large percentage of starch, so as to make them unfit for use in this disease. Recently, however, there has been introduced into the market a Gluten,[*] *not a Gluten flour*, that contains but a mere trace of starch; this is much preferable to all the gluten flours. In very grave cases the bread may be directed made from this Gluten alone, and as the case improves wheat flour may be added thereto in gradually increasing quantities, so that it may be positively known at any time how much starch is being introduced into the system; and if care and judgment be exercised, the quantity which the patient can assimilate can very easily be ascertained, and never any more introduced than the system can appropriate. The exact quantity can be determined by examination of the urine; after it has once been freed from sugar, and kept so for several weeks, then wheat flour may be added to the Gluten, decreasing the quantity as soon as sugar may be detected in the urine.

As to the best mode of making this Gluten into bread, it will be found that by the use of cream of tartar and bicarbonate of sodium it will be made quite light enough; of course, milk and eggs are always added.

---

[*] This Gluten is manufactured by the Crystal Springs Manufacturing Company, Boston, Mass., and by them given the name, "Poluboskos."

Rheumatism.—In these cases a fruit and vegetable diet will be found most advantageous. Acid fruits, such as lemons, are especially to be recommended. Among the vegetables that are to be preferred in these cases may be cited tomatoes, all kinds of greens, such as lettuce, water-cress, spinach, etc., celery, cabbage, sauer-kraut, cauliflower, string beans, asparagus, salsify, pickles, etc. Especially to be avoided are starchy and saccharine articles of food, such as potatoes, corn-starch, arrowroot, beets, oatmeal, pastry, sweet fruits, malt liquors and wines.

To acid fruits and the vegetables allowed, there may be added beef and mutton in moderation, fish, oysters and eggs.

Gout.—Here a mixed diet will be found best, the nitrogenous (such as meats) and saccharine articles (such as contain sugar) used only in limited amounts. The free use of pure water is of great importance. Especially to be avoided are pastry, malt liquors and sweet wines.

Adiposis or Obesity.—To decrease the amount of fat in these cases the patient may be allowed to eat lean beef, lean mutton, lean ham, chicken, dried beef, game, eggs (only one at a meal), fish of all kinds excepting catfish and salmon, buttermilk. Of vegetables he may eat asparagus, cauliflower, onions, spinach, cabbage without a butter dressing, celery, tomatoes, radishes, lettuce, water-cress and greens of all kinds. Of fruits he may eat oranges, lemons, grapes, cherries, berries of all kinds, peaches and sour apples.

Not more than four ounces of bread must be allowed, unless it be the gluten bread recommended for diabetics. Water, tea and coffee without cream and sugar, may be indulged in but very sparingly.

Especially to be avoided are fats in any form, starchy food, such as rice, oatmeal, hominy, potatoes, pies, puddings and all such food as contains sugar; also malt liquors of all kinds, and lastly as little liquid had better be taken as the patient can possibly get along with.

A. L. A. TOBOLDT, M. D.

822 N. Broad Street, Philadelphia.